Nuclear Micro Reactors

Bahman Zohuri

Nuclear Micro Reactors

Springer

Bahman Zohuri
Galaxy Advanced Engineering
Albuquerque, NM, USA

ISBN 978-3-030-47224-5 ISBN 978-3-030-47225-2 (eBook)
https://doi.org/10.1007/978-3-030-47225-2

This Springer imprint is published by the registered company Springer Nature Switzerland AG
The registered company address is: Gewerbestrasse 11, 6330 Cham, Switzerland

This book is dedicated to my son Sash, my grandson Dariush and my granddaughter Donya.

Preface

A new generation of reactors will start producing power in the next few years. They are comparatively tiny—and may be key to hitting our climate goals for the better, free of carbon emissions and free from greenhouse effects.

For the last 20 years, the future of nuclear power has stood in a high bay laboratory tucked away on the Oregon State University campus in the western part of the state. Operated by NuScale Power in the form of Small Modular Reactors (SMR), an Oregon-based energy startup, this prototype reactor represents a new chapter in the conflict-ridden, politically bedeviled saga of nuclear power plants. Or even old companies such as Westinghouse with many years of experience in nuclear power plant in the form of Generation III and now with introduction of transportable Nuclear Micro Reactor eVinci, which has both space exploration into terrestrial domain and military application for a mobile brigade for a rapid deployment process.

NuScale's reactor will not need massive cooling towers or sprawling emergency zones. It can be built in a factory and shipped to any location, no matter how remote due to its modulization technical approach, and it is built around old and traditional knowledge of Light Water Reactor technique. Extensive simulations suggest that it can handle almost any emergency without a meltdown. One reason for that is it barely uses any nuclear fuel—at least compared with existing reactors.

eVinci Micro Reactor cooling system is designed and its cooling system is based on Advanced Heat Pipe technology which is a very dynamic yet as passive cooling system with most safe way without any meltdown disasters either manmade or natural threats.

NASA's approach with heat pipe cooled of kilopower reactor for space exploration and Mars mission in near future is another application of these small reactors yet big energy source for such application that allows to travel beyond terrestrial space.

This is good news for a planet in the grips of a climate crisis. Nuclear energy gets a bad rap in some environmentalist circles, but many energy experts and policymakers agree that splitting atoms is going to be an indispensable part of decarbonizing the world's electricity. In the United States, nuclear power accounts for about two-thirds of all clean electricity, but the existing reactors are rapidly approaching the end of their regulatory lifetimes. Only two new reactors are under construction in the United States, but they are billions of dollars over budget and years behind schedule.

Enter the small modular reactor designed to allow several reactors to be combined into one unit. Need a modest amount of energy? Install just a few modules. Want to fuel a sprawling city? Tack on several more. Coming up with a suitable power plant for a wide range of situations becomes that much easier. As they are small, these reactors can be mass-produced and shipped to any location in a handful of pieces. Perhaps most importantly, small modular reactors can take advantage of several cooling and safety mechanisms unavailable to their big brothers, which all but guarantee they will not become the next Chernobyl or Fukushima.

Nuclear reactors are getting smaller and this is opening up some big opportunities for the industry. A handful of micro reactor designs are under development in the United States, and they could be ready to roll out within the next decade.

These plug-and-play reactors will be small enough to transport by truck and could help solve energy challenges in a number of areas, ranging from remote commercial or residential locations to military bases.

The devastating impacts of climate change caused by burning fossil fuels are forcing countries around the world to look for zero-emissions alternatives for generating electricity.

One such alternative is nuclear energy, and the International Energy Agency—a group focused on energy security, development, and environmental sustainability for 30-member countries—says the transition to a cleaner energy system will be drastically harder without it.

Canada's government appears to be on board, saying nuclear innovation plays a "critical role" in reducing greenhouse gas emissions as Canada moves toward a low-carbon future.

While Canada Deuterium Uranium (CANDU) reactors, a Canadian design, have powered some Canadian communities for decades, the government is now eyeing technology of a different scale. The federal government describes small modular reactors (SMR), as the "next wave of innovation" in nuclear energy technology and an "important technology opportunity for Canada."

In this book, we cover a summary and overall aspect of Generation IV (GEN-IV), or they are also known as Small Modular Reactors (SMRs) as well. In this book, we also cover Nuclear Micro Reactor and its need and implementation within Department of Defense (DOD) military organizations.

Here is what you need to know about them.

What is a small modular reactor?

Traditional nuclear reactors used in Canada can typically generate about 800 MW of electricity, or about enough to power about 600,000 homes at once (assuming that 1 MW can power about 750 homes).

The International Atomic Energy Agency (IAEA), the UN organization for nuclear cooperation, considers a nuclear reactor to be "small" if it generates under 300 MW.

Albuquerque, NM 2016

Bahman Zohuri

Acknowledgment

I am indebted to the many people who aided me, encouraged me, and supported me beyond my expectations. Some are not around to see the results of their encouragement in the production of this book, yet I hope they know of my deepest appreciations. I especially want to thank all my friends, to whom I am deeply indebted, have continuously given their support without hesitation. They have always kept me going in the right direction, specially a true friend Dr. Patrick J. McDaniel.

Above all, I offer very special thanks to my late mother and father, and to my children, in particular, my son Sasha, who always encouraged me while we had in this world for a short time. They have provided constant interest and encouragement, without which this book would not have been written. Their patience with my many absences from home and long hours in front of the computer to prepare the manuscript are especially appreciated.

Contents

About the Author

Bahman Zohuri currently works for Galaxy Advanced Engineering, Inc., a consulting firm that he started in 1991 when he left both the semiconductor and defense industries after many years working as a chief scientist. After graduating from the University of Illinois in the field of physics, applied mathematics, then he went to the University of New Mexico, where he studied nuclear engineering and mechanical engineering. He joined Westinghouse Electric Corporation, where he performed thermal hydraulic analysis and studied natural circulation in an inherent shutdown heat removal system (ISHRS) in the core of a liquid metal fast breeder reactor (LMFBR) as a secondary fully inherent shutdown system for secondary loop heat exchange. All these designs were used in nuclear safety and reliability engineering for a self-actuated shutdown system. He designed a mercury heat pipe and electromagnetic pumps for large pool concepts of an LMFBR for heat rejection purposes for this reactor around 1978, when he received a patent for it. He was subsequently transferred to the defense division of Westinghouse, where he oversaw dynamic analysis and methods of launching and controlling MX missiles from canisters. The results were applied to MX launch seal performance and muzzle blast phenomena analysis (i.e., missile vibration and hydrodynamic shock formation). Dr. Zohuri was also involved in analytical calculations and computations in the study of nonlinear ion waves in rarefying plasma. The results were applied to the propagation of the so-called soliton waves and the resulting charge collector traces in the rarefaction characterization of the corona of laser-irradiated target pellets. As part of his graduate research work at Argonne National Laboratory, he performed computations and programming of multi-exchange integrals in surface physics and solid-state physics. He earned various patents in areas such as diffusion processes and diffusion furnace design while working as a senior process engineer at various semiconductor companies, such as Intel Corp., Varian Medical Systems, and National Semiconductor Corporation. He later joined Lockheed Martin Missile and Aerospace Corporation as Senior Chief Scientist and oversaw research and development (R&D) and the study of the vulnerability, survivability, and both radiation and laser hardening of different components of the Strategic Defense Initiative, known as Star Wars.

This included payloads (i.e., IR sensor) for the Defense Support Program, the Boost Surveillance and Tracking System, and Space Surveillance and Tracking Satellite against laser and nuclear threats. While at Lockheed Martin, he also performed analyses of laser beam characteristics and nuclear radiation interactions with materials, transient radiation effects in electronics, electromagnetic pulses, system-generated electromagnetic pulses, single-event upset, blast, thermo-mechanical, hardness assurance, maintenance, and device technology.

He spent several years as a consultant at Galaxy Advanced Engineering serving Sandia National Laboratories, where he supported the development of operational hazard assessments for the Air Force Safety Center in collaboration with other researchers and third parties. Ultimately, the results were included in Air Force Instructions issued specifically for directed energy weapons operational safety. He completed the first version of a comprehensive library of detailed laser tools for air-borne lasers, advanced tactical lasers, tactical high-energy lasers, and mobile/tactical high-energy lasers, for example.

He also oversaw SDI computer programs, in connection with Battle Management C^3I and artificial intelligence, and autonomous systems. He is the author of several publications and holds several patents.

Chapter 1
Nuclear Micro Reactors: The Next Wave of Innovation

1.1 Introduction

Growth of population globally has direct impact on demand for energy. Almost 18% growth in population and their required daily life on energy and electricity demand presents a different dimension for production of electricity not only from renewable perspective, but also puts nuclear energy resource in different category. New generation of nuclear reactors in the form of Small Modular Reactors (SMRs) or GEN-IV. With new safety factors built into these reactors, with better thermal efficiency output with innovative approach to Combined Cycle (CC) makes them more cost-effective from Return On Investment (ROI) point of view [1–3].

Furthermore, the presence of new renewable technology and suggested solutions by expert in the field for source of energy and energy storage does not eliminate a demand and need for both present and near-term *Nuclear Fission Reactors* in the form of GEN-III (i.e., present) to GEN-IV (i.e., next generation of SMRs) to Nuclear Fusion Reactors in far term.

The rule of thumb for generating electricity is falling into the following category. The requirement for production of electricity is that the electricity generation rate at all times equals the demand for electricity. Economically achieving this goal is easy with fossil fuels because the primary cost of producing electricity is the cost of the fuel, not the cost of the power plant. It is economically viable to operate a fossil plant at part load. As a consequence, in the USA and much of the world the preferred fossil-fuel generating technology is the Gas Turbine Combined Cycle (GTCC)—a low cost machine with rapid response to variable electricity demand with heat-to-electricity efficiencies above 60% [1, 2].

The major growth in the electricity production industry in the last 30 years has centered on the expansion of natural gas power plants based on gas turbine cycles. The most popular extension of the simple Brayton gas turbine has been the combined cycle power plant with the air-Brayton cycle serving as the topping cycle and

B. Zohuri, *Nuclear Micro Reactors*, https://doi.org/10.1007/978-3-030-47225-2_1

the Steam-Rankine cycle serving as the bottoming cycle for new generation of nuclear power plants that are known as GEN-IV. The air-Brayton cycle is an open air cycle and the Steam-Rankine cycle is a closed cycle. The air-Brayton cycle for a natural gas-driven power plant must be an open cycle, where the air is drawn in from the environment and exhausted with the products of combustion to the environment. This technique is suggested as an innovative approach to GEN-IV nuclear power plants in the form and type of Small Modular Reactors (SMRs). The hot exhaust from the air-Brayton cycle passes through a Heat Recovery Steam Generator (HSRG) prior to exhausting to the environment in a combined cycle. The HRSG serves the same purpose as a boiler for the conventional Steam-Rankine cycle [4].

Given the climate change is real fact and low-carbon environment is a mandatory reality, a quest for a new source energy that provides electricity at zero carbon generation becomes a necessity. Thus, our choice of nuclear energy in the form of either fission in near term or fusion in long term is there. In Chap. 2 of this book, we discuss topic of "Why We Need Nuclear Power Plants" based new innovative techniques to make them more efficient as well as safety point of view and cost-effectiveness. Safety aspect of operational version of generation four (GEN-IV) of these reactors in the form of SMRs are number one priority of owners of these reactors given the aftermath of events such as Fukushima Daiichi nuclear disaster (2011), the Chernobyl disaster (1986), the Three Mile Island accident (1979), and the SL-1 accident (1961) are few we can name [5].

The devastating impacts of climate change caused by burning fossil fuels are forcing countries around the world to look for zero-emissions alternatives for generating electricity.

One such alternative is nuclear energy as clean source of energy that is free of carbon dioxide or monoxide generation, and the International Energy Agency (IEA)—a group focused on energy security, development, and environmental sustainability for 30-member countries—says the transition to a cleaner energy system will be drastically harder without it [6].

As part of our need for clean source of energy in the form of nuclear power reactors, NuScale is one of the Small Modular Reactor companies whose designs are going through pre-licensing approval with Canada's nuclear regulator. Many are designed to be small enough to transport by truck or by shipping container. (NuScale Power) and their reactors are so transportable as illustrated in Fig. 1.1, while it is modular as well using technology such as Light Water Reactor (LWR) for licensing purpose through US Nuclear Regulatory Commission (NRC).

Canada's government appears to be on board, saying nuclear innovation plays a "critical role" in reducing greenhouse gas emissions as Canada moves toward a low-carbon future.

While husky CANada Deuterium Uranium (CANDU) reactors (i.e., Fig. 1.2) have powered some Canadian communities for decades, governments are now eyeing technology of a different scale. The federal government describes small modular reactors (SMR) (i.e., see next section), as the "next wave of innovation" in nuclear energy technology and an "important technology opportunity for Canada."

Fig. 1.1 A conceptual illustration of Small Modular Reactor on a truck

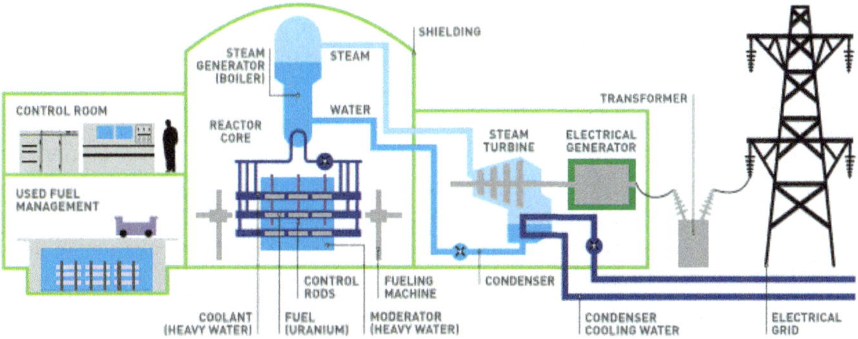

Fig. 1.2 CANDU reactor schematic. (Source: Canadian Nuclear FAQ)

There are currently 18 CANDU reactors in operation in Canada: 8 at Bruce Power, 6 in Pickering, 3 in Darlington, and 1 in Point Lepreau.

CANDU reactors are unique in that they use natural, unenriched uranium as a fuel; with some modification, they can also use enriched uranium, mixed fuels, and even thorium. Thus, CANDU reactors are ideally suited for using material from decommissioned nuclear weapons as fuel, helping to reduce global arsenals.

CANDU reactors are exceptionally safe. The safety systems are independent from the rest of the plant, and each key safety component has three backups. Not only does this redundancy increase the overall safety of the system, but it also makes it possible to test the safety system while the reactor is operating under full power.

1.2 Canada Deployment of Next Generation of Nuclear Technology

In a news released in November 7, 2018, Canada, Poised to Lead the Deployment of Next-Generation Nuclear Technology. The Government of Canada recognizes that innovation in the nuclear sector plays a critical role in reducing greenhouse gas emissions and delivering good, middle-class jobs as Canada moves toward a low-carbon future.

The Honorable Amarjeet Sohi, Canada's Minister of Natural Resources, in 2018 welcomed the release of the Canadian Small Modular Reactor Roadmap, which he described as an "important technology opportunity for Canada, both at home and on the world stage."

He went on to say that small modular reactors represent a promising area of energy innovation in Canada. The roadmap includes recommendations that will help inform ongoing collaboration among federal, provincial, and territorial governments—as well as other stakeholders and Indigenous communities—to ensure Canada becomes a global leader in the development of this new technology.

Small Modular Reactors (SMRs) represent the next wave of innovation in nuclear energy technology. SMRs are designed to be built at a smaller scale than traditional nuclear reactors, with lower upfront capital costs and enhanced safety features. They have potential to provide non-emitting energy in a wide range of applications, such as grid-scale electricity generation and heavy industry, including in remote communities.

The Department of Natural Resources convened interested provinces, territories, power utilities, Indigenous communities, and other stakeholders to support the development of the roadmap. The roadmap is a result of a 10-month engagement process with the industry and potential end-users, including Indigenous and northern communities and heavy industry. It includes over 50 recommendations in areas such as waste management, regulatory readiness, and international engagement. It also highlights the need for ongoing engagement with civil society, northern and Indigenous communities, and environmental organizations. The roadmap arose out of last year's Generation Energy consultation process—the largest national conversation about energy in Canada's history.

The Government of Canada welcomes the Canadian Small Modular Reactor Roadmap and is presently reviewing its recommendations.

1.3 What is a Small Modular Reactor?

Based on augmentation of Nuclear Power Plant in Canada that are operating on their electric grid, a traditional nuclear reactors used in Canada can typically generate about 800 MW of electricity, or about enough to power about 600,000 homes at once (assuming that 1 MW can power about 750 homes). Similarly, in the USA the largest operational reactor on line is Palo Verde near Tonopah, Arizona, that is producing 1311 MW power generation per unit with total of three units as illustrated in Fig. 1.3.

The Palo Verde Generating Station is the largest power plant in the USA by net power generation. Its average electric power production is about 3.3 gigawatts (GW), and this power serves about 4 million people. The Arizona Public Service Company (APS) operates and owns 29.1% of the plant. Its other major owners include the Salt River Project (SRP) (17.5%), the El Paso Electric Company (15.8%), Southern California Edison (SCE) (15.8%), PNM Resources (10.2%), the Southern California Public Power Authority (SCPPA) (5.9%), and the Los Angeles Department of Water and Power (5.7%).

Currently, all large commercial power reactors in the USA and most in the rest of the world are based on "light water" designs—that is, they use uranium fuel and ordinary water for cooling. By contrast, an emerging class of small reactors come in widely varying designs and use a variety of fuels and cooling systems, some can even utilize existing legacy radioactive waste as a fuel source. They range from downsized light water reactors to more exotic liquid metal-cooled fast reactors, with the smallest designs beginning at a 10 MW capacity.

Moreover, the International Atomic Energy Agency (IAEA), the UN organization for nuclear cooperation, considers a nuclear reactor to be "small" if it generates under 300 MW.

Designs for small reactors ranging from just 3 to 300 MW have been submitted to Canada's nuclear regulator, the Canadian Nuclear Safety Commission (CNSC), for review as part of a pre-licensing process.

Fig. 1.3 Aerial photo of Palo Verde Generation Station

Thus, such reactors are considered "modular" because they are designed to work either independently or as modules in a bigger complex (as is already the case with traditional, larger reactors at most Canadian nuclear power plants). A power plant could be expanded incrementally by adding additional modules. See Fig. 1.4.

Modules are generally designed to be small enough to make in a factory and be transported easily—for example, via a standard shipping container. See Fig. 1.5.

Bear in mind that small modular reactors are nuclear energy's future and scaled-down nuclear power plants offer price gains over conventional sites such as Palo Verde power generation station in Arizona.

As delays mount at large new nuclear power projects around the world, more attention is turning to smaller alternatives, which industry experts hope may help provide the next generation of electricity.

The so-called small modular reactors—miniature nuclear power plants with a capacity of less than 300 MW—could provide an alternative to mega-plants like the two 1.6 GW reactors planned at Hinkley Point in Somerset.

The UK project is one of a number of delayed or abandoned nuclear power schemes, which have left policymakers around the world looking for cheaper, less risky options to meet electricity demand.

SMRs are designed as shrunken versions of larger plants; they can be made in factories and moved by train, truck, or barge to the site. Developers say that if enough are built in the same factory, costs per unit of energy output can be driven down well below those of larger plants.

Small reactors are already used on nuclear submarines and in some developing countries such as India and Pakistan. But only recently have the industry and politicians begun to take seriously the idea that they could be made economically on a large scale.

Small Modular Reactors (SMRs) from Total Cost of Ownership (TCO) and Return On Investment (ROI) promise all the benefits of nuclear—low cost and green power—but without the significant cost and schedule overrun issues associated with

Fig. 1.4 Modular reactor concept. (Source: Generation IV International Forum)

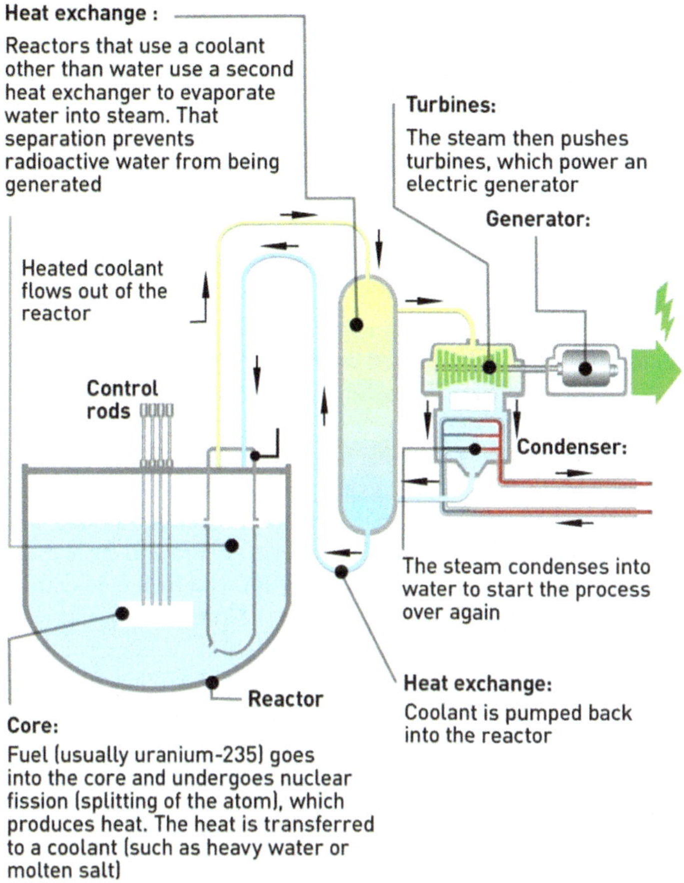

Fig. 1.5 Small Modular Reactors are nuclear energy's future in transit

large-scale size power plant such as traditional one of GEN III, where the beset conventional large nuclear projects have to bear.

Since the invention of nuclear power, bigger has generally been seen to be better. Once a company had gone through the time and expense of securing a site along with planning approval and grid connections, most wanted to build as much capacity on that site as possible [7].

However, many of those stations have been plagued with problems, which some blame on their size. Plans by EDF, the French energy company, to build new reactors in France and Finland, for example, have gone billions of euros over budget—something many experts blame on the difficulty of making such large structures safe. Furthermore, building the large-scale Nuclear Power Plants (NPPs) take more time to build them, simply because there are more huge structures to deal with and to protect against natural and man-made disaster and so many other safety systems that provide safety protection against any terrorist acts toward these plants.

Large projects such as these have also had trouble getting financed—one of the principal causes of delay at Hinkley Point has been the difficulty EDF is having in raising the money needed for the £18 billion project.

For now, small-scale nuclear industry proponents are focused on proving the technology can work at costs low enough to make it competitive. The countries that are furthest along are, unsurprisingly, those with the most developed nuclear energy industries.

Russia is in the process of converting two small reactors which used to power icebreakers. They will eventually be placed on barges which can then be moved to where they are needed [7].

All these falls in our thinking of holding the costs down—as long as enough SMRs are manufactured and deployed for commissioning at give sites. Figures 1.6 and 1.7 shows such artistic renderings of the Small Modular Reactor (SMR) site in comparison to a traditional one.

The USA and the UK are both trying to catch up. The UK recently took a leaf out of the US book when it announced it would run a competition to find the best Small Modular Reactor (SMR) design, with £250 million on offer to help with research and development. From combined cycle and open air-Brayton cycle thermodynamics point of view advanced version of such SMRs with high temperature will provide better thermal efficiency output does bring the Total Cost of Ownership (TCO) and Return On Investment (ROI) to a very reasonable cost-effectiveness [1–4].

"The US and the UK are in a race at the moment, and that is driving both forward," says Jared DeMeritt, program director of MPower, an SMR developer. "We think 2025 is a realistic start date for the first small modular reactor in the west, which will be in one of these two countries." This is according to the UK Financial Times [7].

MPower's design shows some of the ways that smaller plants can avoid the pitfalls of larger ones. In its case, MPower plans to bury all safety-critical equipment—including the reactor and the fuel vessels—underground, thereby minimizing the need for expensive physical defenses [7].

Fig. 1.6 Traditional nuclear plant real estate versus SMR footprint

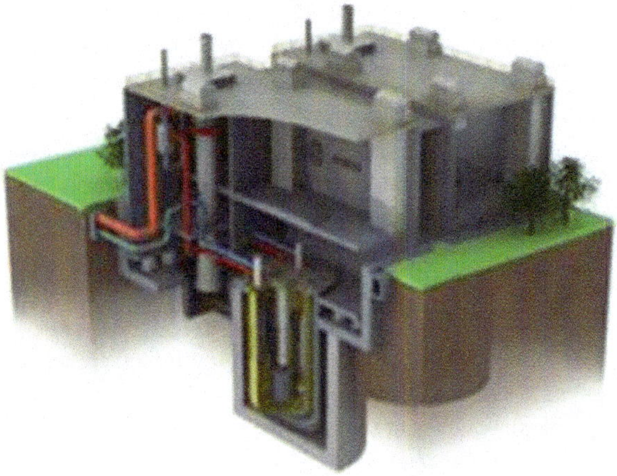

Fig. 1.7 Artistic layout of SMR installation

Despite the optimism among some in the industry, there remain significant hurdles to widespread use of SMRs. Firstly, even those building them privately admit the first ones will cost roughly the same per unit of electricity produced by a large reactor until costs can be driven down. One executive says: "Over time, we think we can get the costs down—as long as enough of them are commissioned" [7].

But advocates of SMRs say that even if they prove more expensive for the electricity produced, costs are less likely to escalate and more likely to be fully funded [7].

David Hess of the World Nuclear Association says: "Financing is a huge policy risk, and SMRs reduce that. And if the project goes wrong, at least less money has been wasted" [7].

1.4 Nuclear Reactors Driving Electricity Generation

In this section, we answer the question about "*How do nuclear reactors generate electricity*?". The basic cycle is as described below.

Nuclear reactors of all sizes are powered by nuclear fission—the process of splitting atoms of nuclear fuel, typically uranium, into smaller atoms. That generates heat. This definition applies to present situation where fission reaction is driving the Nuclear Power Plant (NPP) at present time. However, in near future fusion reaction also will be source of energy to drive the Nuclear Power Plant (NPP) of different type based on Magnetic Confinement Fusion (MCF) [8] or the other possibility will be Inertial Confinement Fusion (ICF) [9] as well. Both type of these rectors are thermonuclear-driven fusion kinds rather than fission-driven one.

In thermal power plants, the heat turns water into steam, and the steam pushes turbines that generate electricity. That part of the process is the same whether the heat is generated by nuclear power, burning fossil fuels such as coal or natural gas, or even concentrated solar energy. See Fig. 1.8 for this concept.

What every country that is involved with technology of Small Modular Reactors (SMRs) are in agreement with each other in respect to the advantages of SMRs Nuclear Power Plant (NPP) over the traditional one.

For example, the Canada's government says SMRs are designated to have lower upfront capital costs and enhanced safety features compared to traditional reactors.

Because of their small size, most could be completely built in a factory and installed module by module, making construction quicker, more efficient and theoretically cheaper, according to the World Nuclear Association [10], which represents the nuclear industry. Upfront costs, especially, would be lower, since modules could be added as needed, when the demand for electricity rises, instead of being paid for all at once.

The US support for Small Modular Reactors (SMRs) through Department of Energy (DOE) falls back to January of 2012, when DOE called for applications from industry to support the development of one or two US Light Water Reactor (LWR) designs, allocating $452 million over 5 years period through the SMR Licensing Technical Support (LTS) program. Thus, four applications were suggested and made, from Westinghouse (W), Babcock & Wilcox (BW), Holtech, and NuScale Power companies and the units of each module were ranging from 225 MWe down to 45 MWe.

In March 2012, the DOE signed agreements with three companies interested in constructing demonstration small reactors at its Savannah River site in South Carolina. The three companies and reactors are: Hyperion (now GEN IV Energy) with a 25 MWe fast reactor, Holtec with a 160 MWe PWR, and NuScale with its 45 MWe PWR (since increased to 60 MWe). The agreements concerned the provision of land but not finance. The DOE was in discussion with four further small

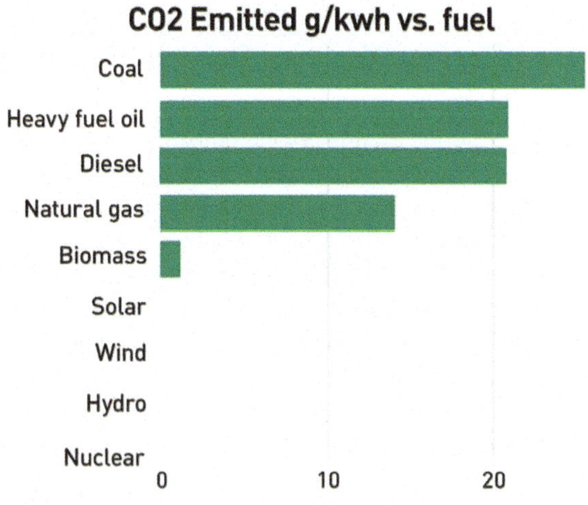

Fig. 1.8 Heat generated source. (Source: Government of Canada)

reactor developers regarding similar arrangements, aiming to have in 10–15 years a suite of small reactors providing power for the DOE complex. (Over 1953–1991, Savannah River was where a number of production reactors for weapons plutonium and tritium were built and run.)

SMR Start has called for the DOE's LTS program for SMRs to be extended to 2025 with an increase in funding. It pointed out: "Private companies and DOE have invested over $1 billion in the development of SMRs. However, more investment, through public–private partnerships is needed in order to assure that SMRs are a viable option in the mid-2020s. In addition to accomplishing the public benefit from SMR deployment, the federal government would receive a return on investment through taxes associated with investment, job creation and economic output over the lifetime of the SMR facilities that would otherwise not exist without the US government's investment."

Canadian support for SMRs technology goes back to June of 2016, when it was announced through a report for the Ontario Ministry of Energy that was focused on nine designs under 25 MWe per module for off-grid remote sites. All had a medium level of technology readiness and were expected to be competitive against diesel. Two designs were integral Pressurized Water Reactors (PWRs) of 6.4 and 9 MWe, three were High Temperature Reactors (HTRs) of 5, 8, and 16 MWe, two were Sodium-cooled Fast Reactors (SFRs) (i.e., Fig. 1.9) of 1.5/2.8 and 10 MWe, one was

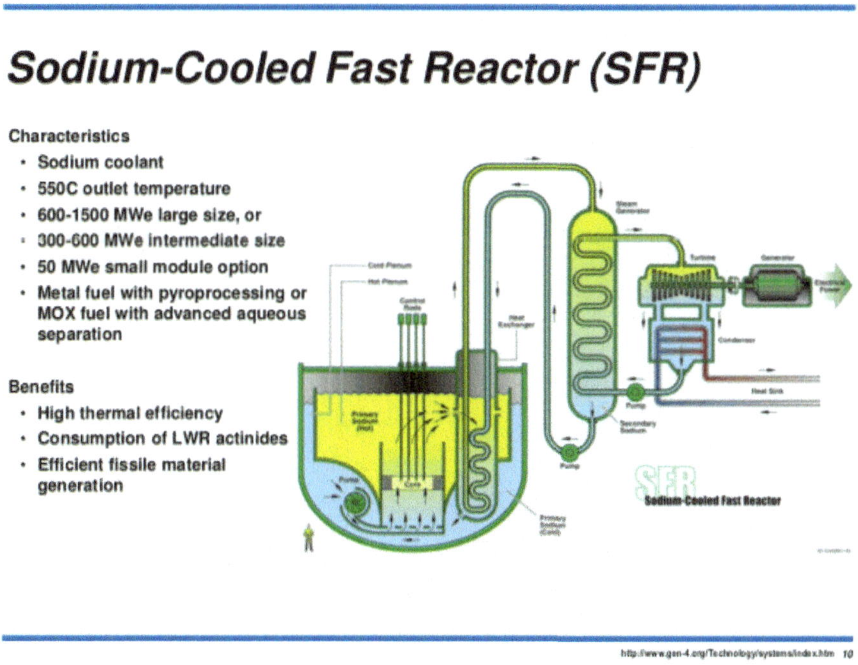

Fig. 1.9 Sodium-cooled Fast Reactor (SFR). (Source: www.wikipedia.com)

a Lead-cooled Fast Reactor (LFR) (i.e., see Fig. 1.10) of 3–10 MWe, and one was an MSR of 32.5 MWe. Four were under 5 MWe (an SFR, LFR, and two HTRs).

Note that a Sodium-cooled Fast Reactor (SFR) is a fast neutron reactor cooled by liquid sodium.

The acronym SFR particularly refers to two Generation IV reactor proposals, one based on existing Liquid Metal Fast Breeder Reactor (LMFR) technology using Mixed Oxide (MOX) fuel [11], the other based on the metal-fueled integral fast reactor.

Several sodium-cooled fast reactors have been built, some still in operation, and others are in planning or under construction.

Ontario distinguishes "grid scale" SMRs above 25 MWe from these very small-scale reactors.

Lead-cooled Fast Reactors (LFRs) are fast spectrum reactors cooled by molten lead (or lead-based alloys) operating at high temperatures and at near atmospheric pressure, conditions enabled because of the very high boiling point of the coolant (up to 1743 °C) and its low vapor pressure. Due to the fundamental thermodynamic and neutronic characteristics of lead as a coolant, LFRs offer great potential for new reactor designs that achieve a high degree of inherent safety, simplified operation, and excellent economic performance while providing the fuel material management advantages characteristic of fast reactors. LFR designs span the range of reactor sizes and potential deployment scenarios.

Pushing the physical sizes of the new generation (i.e., GEN IV) from small to smaller, Westinghouse Nuclear Micro Reactor heat Pipe design suggests a new innovative technical approach as illustrated in Fig. 1.11.

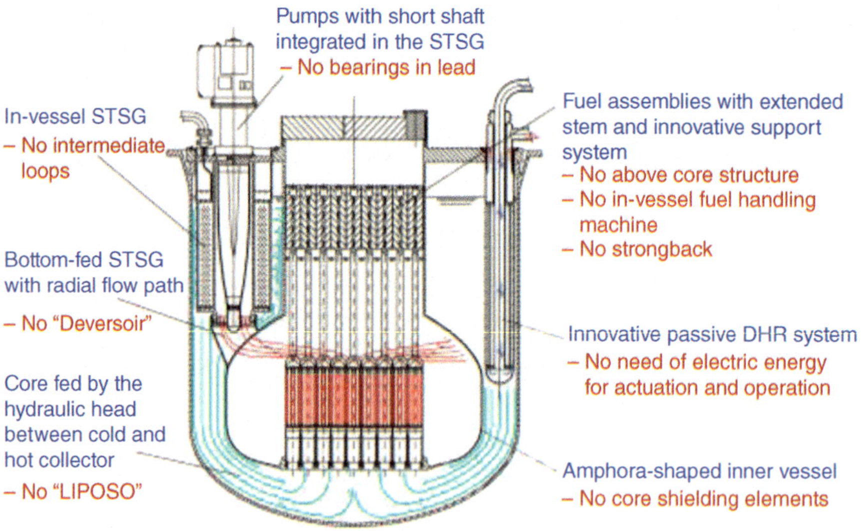

Fig. 1.10 Lead-cooled Fast Reactor (LFR)

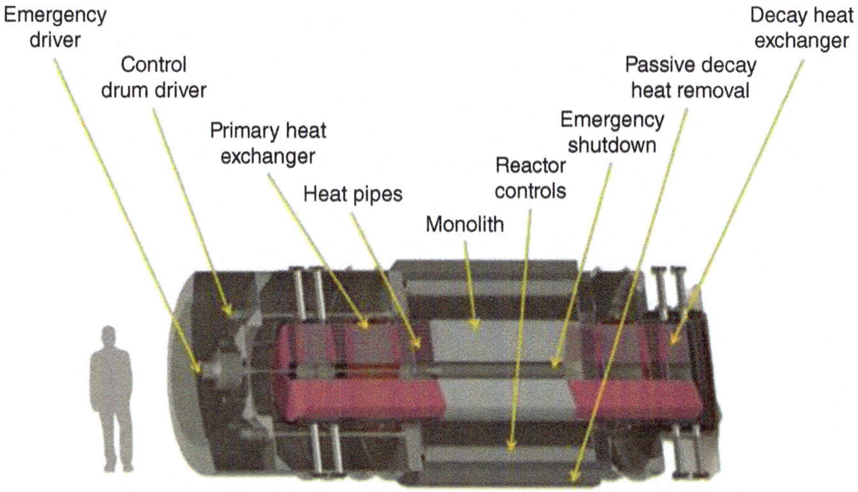

Fig. 1.11 Westinghouse eVinci heat pipe nuclear micro reactor layout. (Source: www.wikipedia.com)

Distinct from other small reactor designs, the eVinci is a heat pipe reactor, using a fluid in numerous sealed horizontal steel heat pipes to conduct heat from the hot fuel (where the fluid vaporizes) to the external condenser (where the fluid releases latent heat of vaporization) with heat exchanger. No pumps are needed to effect continuous isothermal vapor/liquid internal flow at low pressure. The principle is well established on a small scale, but here a liquid metal is used as the fluid and reactor sizes up to several megawatts are envisaged. Experimental work on heat pipe reactors for space has been with much smaller units (about 100 kWe), using sodium as the fluid. They have been developed since 1994 as a robust and low technical risk system for space exploration with an emphasis on high reliability and safety.

The eVinci reactors would be fully factory built and fueled. As well as power generation, process heat to 600 °C would be available. Units would have 5–10-year operational lifetime, with walkaway safety due to inherent feedback diminishing the nuclear reaction with excess heat, also effecting load following.

Bear in your mind that heat pipe is passive heat transfer devise that holds no moving part in it and does not require to take on task of heat transfer [12, 13].

The US Nuclear Regulatory Commission (NRC) has released a draft white paper on its strategy for reviewing licensing applications for advanced non-light water reactor technologies. The NRC said it expects to finalize the draft paper by November, with submission of the first non-Light Water Reactor (non-LWR) application expected by December 2019. By mid-2019 the NRC had been formally notified by six reactor designers of their intention to seek design approval. These included three MSRs, one High Temperature Reactor (HTR), one Fast Neutron Reactor (FNR), and the Westinghouse eVinci heat pipe reactor [10].

Note that a Fast-Neutron Reactor (FNR) or simply a fast reactor is a category of nuclear reactor in which the fission chain reaction is sustained by fast neutrons (carrying energies above 0.5 MeV or greater, on average), as opposed to thermal neutrons used in thermal-neutron reactors.

Among six types of Generation IV (GEN-IV) reactors under conceptual study for types of Small Modular Reactor (SMR) is the Molten Salt Reactors (MSRs) that shows a promising future for build and going to production. See Fig. 1.12.

A Molten Salt Reactor (MSR) is a class of nuclear fission reactor in which the primary nuclear reactor coolant and/or the fuel is a molten salt mixture. MSRs offer multiple advantages over conventional nuclear power plants, although for historical reasons, they have not been deployed, yet is under research study.

The concept was first established in the 1950s. The early Aircraft Reactor Experiment was primarily motivated by the small size that the technique offered, while the Molten Salt Reactor Experiment was a prototype for a thorium fuel cycle breeder nuclear power plant. The increased research into Generation IV reactor designs renewed interest in the technology.

With what we have seen so far the nuclear power plant manufacturing companies and industry are making a big bet on small power plant with NuScale (i.e., Fig. 1.13) in the lead following with Westinghouse (i.e., Fig. 1.11) and Holmic small reactor (i.e., Fig. 1.14) behind them, while General Electric Hitachi Collaborating with ARC in same space (i.e., Fig. 1.15).

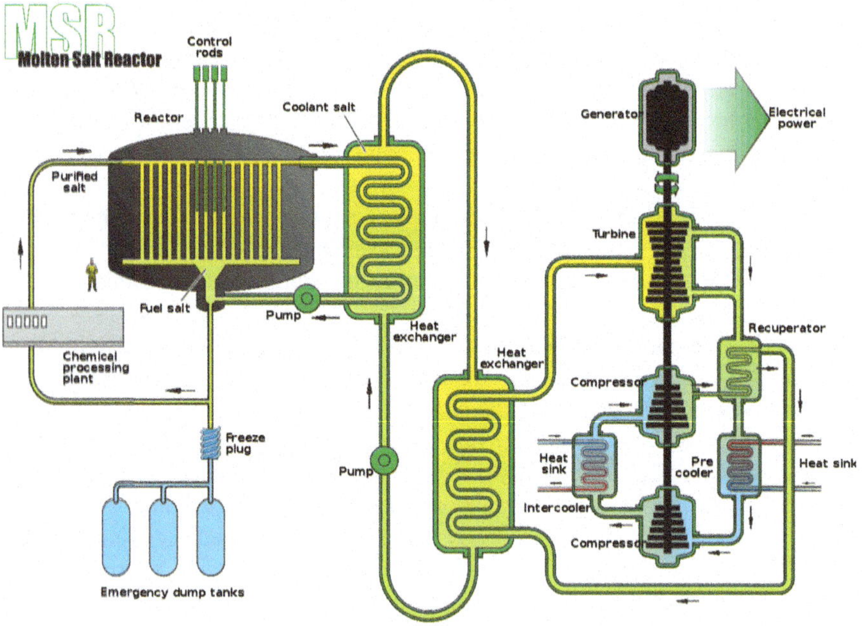

Fig. 1.12 Example of a Molten Salt Reactor scheme. (Source: www.wikipedia.com)

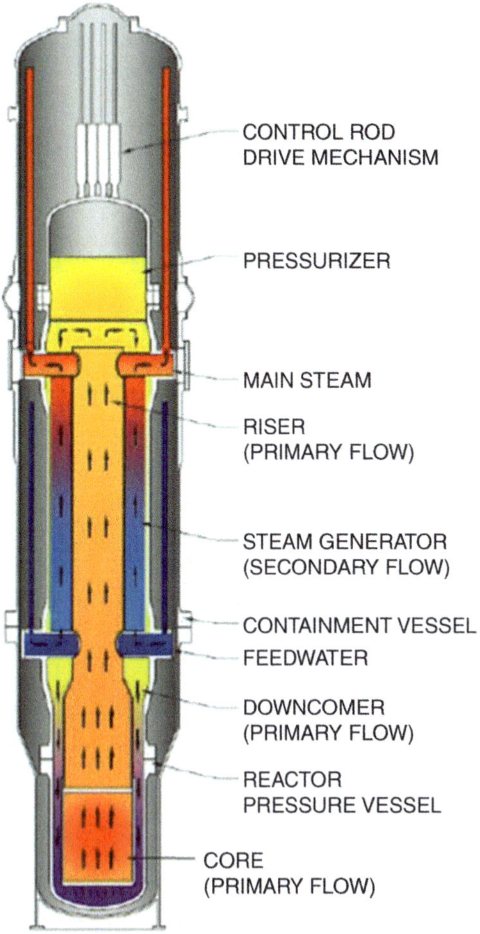

Fig. 1.13 NuScale small modular reactor configuration. (Source: www.wikipedia.com)

Until now, generating nuclear power through Generation III (GEN-III) has required massive facilities surrounded by acres of buildings, electrical infrastructure, roads, parking lots, and more. The nuclear industry is trying to change that picture—by going small, thus smaller footprint from real estate perspective.

Efforts to build the nation's first "advanced small modular reactor," or SMR, in Idaho, are on track for it to become operational by the mid-2020s. The project took a crucial step forward when the company behind it, NuScale, secured an important security certification from the Nuclear Regulatory Commission (NRC).

But worldwide, the first ones could be generating power by 2020 in China, Argentina, and Russia, according to the International Atomic Energy Agency (IAEA).

The debate continues over whether this technology is worth pursuing, but the nuclear industry is not waiting for a verdict. Nor, as an energy scholar presumably. This new generation of smaller and more technologically advanced reactors offer many advantages, including an assembly-line approach to production, vastly reduced

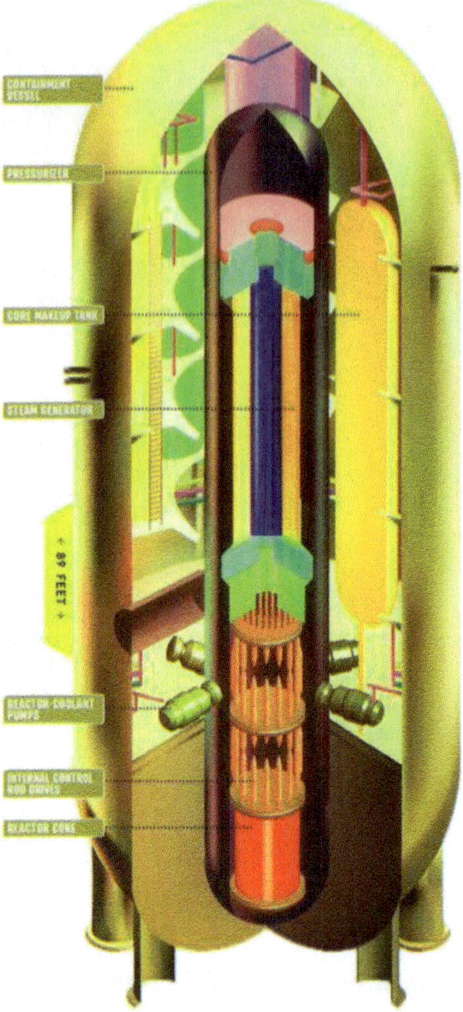

Fig. 1.14 Possible Holtic small reactor configuration. (Source: www.wikipedia.com)

meltdown risks and greater flexibility in terms of where they can be located, among others. See next section of this chapter for trade-off and comparison with tradition nuclear power plant in operation now.

Now the question is how small is small and will be considered as part Small Modular Reactor (SMR) technique or we need to consider something at the size of Nuclear Micro Reactor (NMR).

Most small modular reactors now in the works range between 50 MW—roughly enough power for 60,000 modern US homes—and 200 MW. And there are designs for even smaller such as modular "mini" or modular "micro-reactors" that generate as few as 3–4 MW.

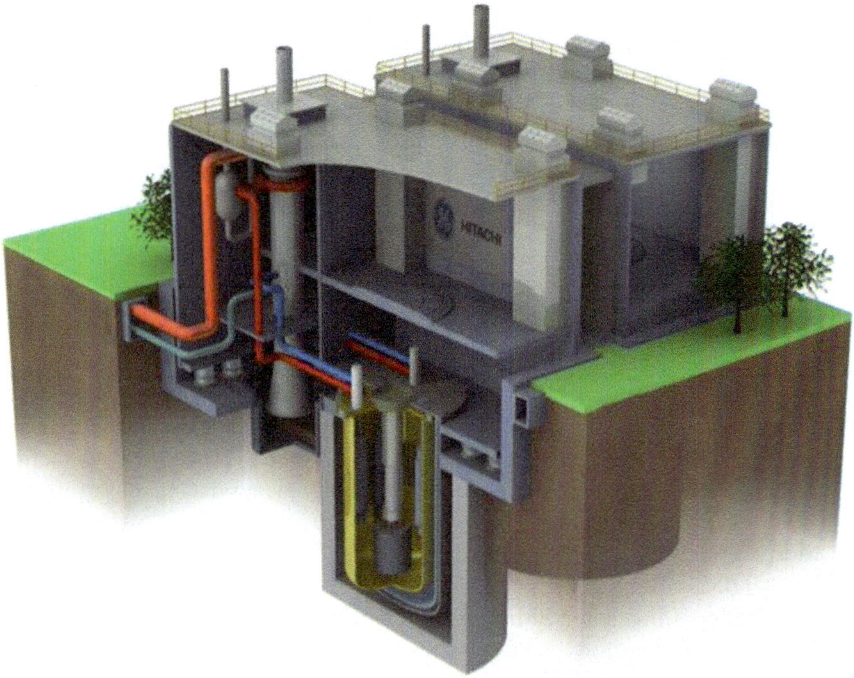

Fig. 1.15 GE Hitachi and ARC small reactor layout. (Source: www.wikipedia.com)

In contrast, full-sized nuclear reactors built today will generate about 1000–1600 MW of electricity, although many built before 1990, including over half the 99 reactors now operating in the USA, are smaller than this.

1.5 What Are the Advantages of SMRs Over Traditional Nuclear Power Plants?

All the scientists and engineers involved in design of Small Modular Reactors (SMRs) all in agreement that such rectors have lower upfront capital consist and enhanced safety features compared to traditional reactors.

Because of their small size, most could be completely built in a factory and installed module by module, making construction quicker, more efficient and theoretically cheaper, according to the World Nuclear Association (WNA), which represents the nuclear industry. Upfront costs, especially, would be lower, since modules could be added as needed instead of being paid for all at once [10].

An additional reason for interest in SMRs is that they can more readily slot into brownfield sites in place of decommissioned coal-fired plants, the units of which are seldom very large—more than 90% are under 500 MWe, and some are under 50

MWe. In the United States, coal-fired units retired over 2010–12 averaged 97 MWe, and those expected to retire over 2015–25 average 145 MWe.

As it was stated in above by definition, Small Modular Reactors (SMRs) are defined as nuclear reactors generally 300 MWe equivalent or less, designed with modular technology using module factory fabrication, pursuing economies of series production, and short construction times. This definition, from the World Nuclear Association, is closely based on those from the IAEA and the US Nuclear Energy Institute. Some of the already-operating small reactors mentioned or tabulated below do not fit this definition, but most of those described do fit it.

Another feature that is predicted to lower the cost is that it is easier to cool SMRs because of their larger surface area-to-volume ratio. That means their safety systems do not need to be as complex. Most can rely on "passive" built-in safety features in the event of a malfunction, rather than special systems that need to be activated. In particular those Advanced High Temperature Small Modular Reactors (SMRs) are an excellent candidate for augmenting open air-Brayton cycle for a better thermal efficiency output, thus reduces cost of ownership by producing more efficient energy for generating electricity [1].

Moreover, can we say that are there any other differences between SMRs and the traditional (GEN-III) Nuclear Power Plants?

The answer is that some SMR designs are effectively scaled-down versions of traditional nuclear reactors, such as NuScale SMR design is around the technology of Light Water Reactors (LWR) that makes to obtain operational licensing by far out much easier from Nuclear Regulatory Commission (NRC) perspective, but some also incorporate next-generation nuclear technologies and designs.

For example:

- *Molten salt reactors* that use molten salt—salt that has melted into a liquid at high temperatures—instead of water as a coolant and dissolve the fuel in the salt. That allows them to operate at regular atmospheric pressure instead of the high pressure that traditional reactors operate at.
- *Liquid metal fast reactors* use liquid sodium or lead as a coolant. Both molten salt and liquid metal fast reactors can reuse and consume fuel from other reactors.
- *High temperature gas reactors* use an inert gas such as helium as a coolant and can operate at a higher temperature, making them more efficient.

There are other designs for SMRs nuclear power plant as well and overall total of six of these reactors are under consideration as Generation IV (GEN-IV).

1.6 Small Modular Reactors Applications

Innovative small reactors can help meet clean energy goals and make electricity more accessible for all.

In addition to reducing carbon emissions, SMRs will use a tiny fraction of land due to its small footprint compared to wind and solar. Small reactors can power retired fossil sites, match electricity output with demand, integrate with renewables, and be used for heat, desalination, and other applications.

Due to the relatively small and mobile nature of the SMRs, they are of great interest to governments and private groups alike. Large nuclear reactors are fraught with complications such as finding space for installment, financing, and timely construction. SMRs are a viable solution, and governments such as Russia and France have already begun construction and utilization of this technology [1, 2, 14].

Furthermore, SMRs are of great interest due to their potential to curb CO_2 emissions as an alternative source of power [15]. This report seeks to evaluate the potential benefits and applications for SMRs in the upcoming decade as well as consider the potential downsides of utilizing SMRs moving forward. See Fig. 1.16.

SMRs are already being utilized by multiple parties, including "on nuclear submarines and in some developing countries such as India and Pakistan" [1, 4, 14]. Some major upsides of SMRs compared to larger nuclear reactors is that they are more transportable, require less uranium fuel, which could potentially lead to fewer meltdowns, and are more affordable at initial market prices. One of the major advantages to SMR technology is the initial economic benefit. While large nuclear reactor sites are extremely costly and difficult to finance, SMRs are more feasible and thus open up the opportunity of harnessing nuclear energy for more parties globally. To illus-

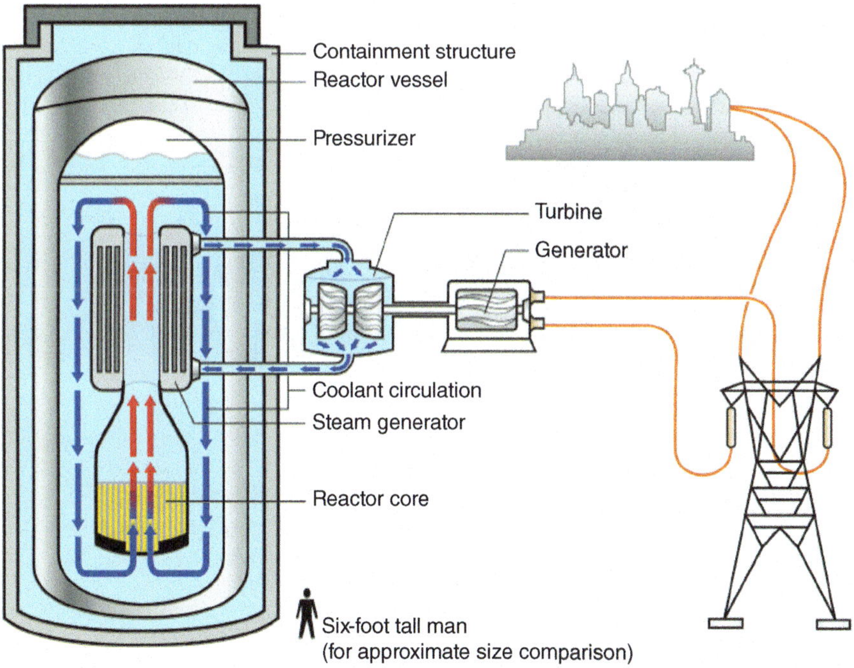

Fig. 1.16 Illustration of a light water small modular nuclear reactor. (Source: GAO, based on Department of Energy documentation, GAO-15-652; Courtesy of the U.S. Government Accountability Office)

trate this matter, consider France for a moment. A French energy company, EDF, had plans to build new large reactor sites in France and Finland. However, due to potential safety concerns, "the plans went billions of euros over budget" [1, 4, 14].

The problem lies in the fact that large nuclear reactors take more time to build and check safety features. Thus, SMR's could help overcome this barrier of financial and time pressures. Evidently, SMRs have already begun to infiltrate our world in tangible ways. However, SMRs are not as widespread as one might predict considering their many advantages. As governments and private entities begin to adopt SMRs and harness the capability of SMRs to produce "cleaner energy," it is also important to consider the potential downsides and dangers associated with SMRs.

In Canada, according to the Canadian Nuclear Safety Commission (CNSC) report, there are three main areas where SMRs could be used as well and they are listed below and illustrated by Fig. 1.17:

- Traditional, on-grid power generation, especially in provinces looking for zero-emission replacements for CO_2-emitting coal plants.
- Remote communities that currently rely on polluting diesel generation.
- Resource extraction sites, such as mining and oil and gas.

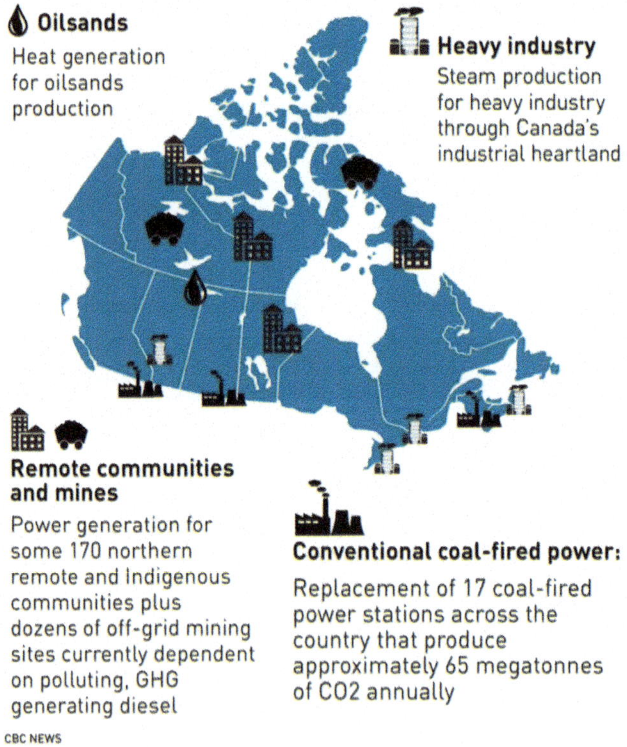

Fig. 1.17 Resource extraction sites illustration. (Source: Organization of Canadian Nuclear Industries)

Naturally, any new innovative technology has its own challenges at the beginning of its introduction to society to be accepted. So we need to ask similar questions about Small Modular Reactors (MSRs) to see "What challenges do SMRs face before they are built?"

While small modular reactors should theoretically be cheaper than traditional reactors, their actual costs will not be known until some designs are actually built and operating, noted Scott Montgomery, an affiliate faculty member at the University of Washington, who lectures and writes about global energy, in an article in The Conversation last June 28, 2018 [16].

He added that while SMRs are designed to produce less nuclear waste than larger reactors, disposal remains an issue.

The World Nuclear Association says licensing costs for an SMR are "potentially a challenge" as they are not necessarily cheaper than they are for a large reactor. However, the Canadian Nuclear Safety Commission (CNSC) notes that licensing costs are a small part of the cost of developing the technology and include many activities that would have to occur anyway to show the technology is reliable and safe as illustrated in Fig. 1.18.

Like the USA, Canada does not yet have a permanent nuclear waste repository although the Nuclear Waste Management Organization (NWMO) is currently working to select a site.

However, Canada is not as close as the USA for the SMS to operate in their backyard. Natural Resources Canada released an "SMR roadmap" in November, with a series of recommendations about regulation readiness and waste management for SMRs.

In Canada, about a dozen companies are currently in pre-licensing with the CNSC, which is reviewing their designs.

In November 2018, Natural Resource Canada (NRC) released a Small Modular Road Roadmap "SMR roadmap" with a series of recommendation about regulation readiness and waste management for SMRs.

In Canada, about a dozen companies are currently in pre-licensing with the Canadian Nuclear Safety Commission (CNSC), which is reviewing their designs.

Ultra-Safe Nuclear's Micro Modular Reactor Energy System is designed to fit in a standard shipping container. The company is partnering with Global First Power and Ontario Power Generation, which are in talks with AECL and CNSC about preparing a site for a reactor at the Chalk River Laboratories. (Ultra-Safe Nuclear) as the conceptual illustration in Fig. 1.18.

The furthest project ahead is one involving Global First Power, in partnership with Ontario Power Generation and Ultra-Safe Nuclear Corp. In April, it began discussions with the Crown corporation Atomic Energy of Canada Ltd. (AECL) and the CNSC about preparing a site for a reactor at AECL's Chalk River Laboratories. There have been plans to have an SMR demonstration plant built at an AECL site by 2026.

According to the International Atomic Energy Agency (IAEA), there are currently four SMRs in advanced stages of construction in Argentina, China, and Russia.

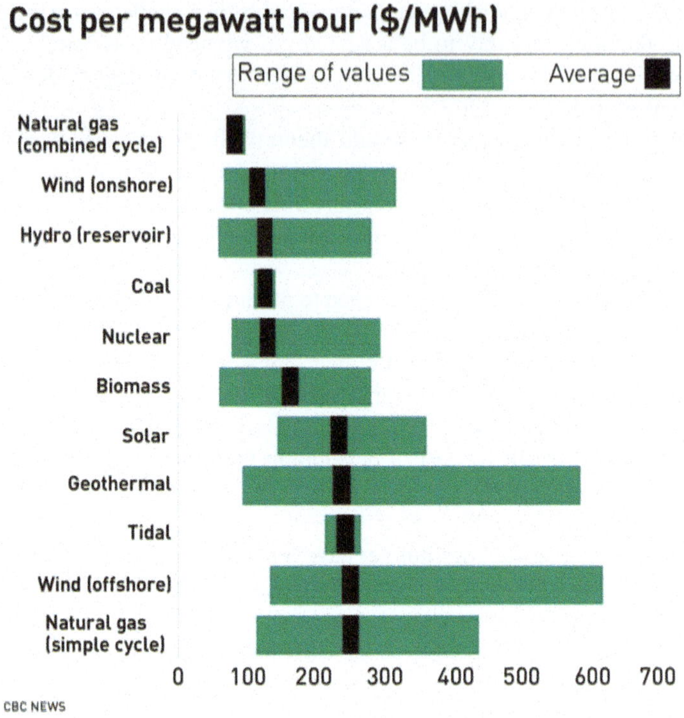

Fig. 1.18 Cost per megawatt hour production. (Source: The National Energy Board)

1.7 Integral Modular Small Modular Reactor

Canada-based Terrestrial Energy set up in 2013 has designed the Integral MSR (IMSR). This simplified MSR integrates the primary reactor components, including primary heat exchangers to secondary clean salt circuit, in a sealed and replaceable core vessel that has a projected life of 7 years. The IMSR will operate at 600–700 °C, which can support many industrial process heat applications. The moderator is a hexagonal arrangement of graphite elements. The fuel-salt is a eutectic of low-enriched uranium fuel (UF_4) and a fluoride carrier salt at atmospheric pressure. Secondary loop coolant salt is ZrF_4-KF. Emergency cooling and residual heat removal are passive. Each plant would have space for two reactors, allowing 7-year changeover, with the used unit removed for offsite reprocessing when it has cooled, and fission products have decayed. Terrestrial Energy hopes to commission its first commercial reactor by the 2020s (Fig. 1.19).

The IMSR is scalable and three sizes were initially presented: 80, 300, and 600 MWt, ranging from 30 to 300 MWe, but from 2016 the company is focused on 400 MWt/192 MWe. The total levelized cost of electricity from the largest is projected to be competitive with natural gas. The smallest is designed for off-grid, remote power applications, and as prototype. Industrial heat at about 600 °C is also envisaged in 2016 plans.

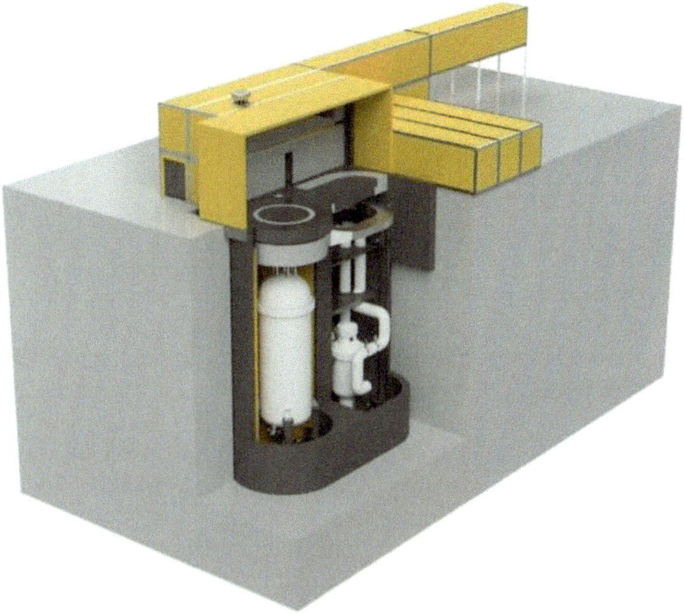

Fig. 1.19 Conceptual layout of nuclear micro reactor

Compared with other MSR designs, the company deliberately avoids using thorium-based fuels or any form of breeding, due to "their additional technical and regulatory complexities."

In November 2017, Terrestrial Energy completed phase 1 of the Canadian Nuclear Safety Commission's (CNSC's) pre-licensing vendor review of the IMSR-400. The company plans to submit either an application seeking design certification or a construction permit application for the IMSR-400 no later than October 2019 to the NRC. It hopes to commission its first commercial reactor in the 2020s.

To meet the increasing energy demands of global prosperity, while protecting the environment and the air in our cities that we breathe, we need a game changer. Beside the USA the Canadian such as Terrestrial Energy is pushing toward Integral Modular Small Reactor with advanced version of these modular reactor technology as illustrated in Fig. 1.20.

Terrestrial Energy is developing revolutionary—not simply evolutionary—nuclear technology. The IMSR® uses completely different nuclear technology for its Advanced Modular Reactor—molten salt fuel instead of conventional solid fuel. With this proven approach, IMSR® Generation IV nuclear power plants are more affordable, cost-competitive, and versatile than conventional nuclear power plants.

IMSR® technology can be brought to market quickly. IMSR® power plants can be built in 4 years and produce electricity or industrial heat at prices competitive with fossil fuels, while emitting no greenhouse gases.

Fig. 1.20 Conceptual advanced modular reactor layout

Globally, other countries like the UK and China are in support of SMRs Nuclear Power Plant (NPP) and other countries through Urenco a group of a nuclear fuel company operating several Uranium enrichment plants in Germany, the Netherlands, the USA, and the UK has called for European development of very small—4 MWe—"Plug and Play" inherently safe reactors based on graphite-moderated High Temperature Reactor (HTR) concepts. See Fig. 1.21.

HTRs can potentially use thorium-based fuels, such as highly enriched or low-enriched uranium with Th, U-233 with Th, and Pu with Th. Most of the experience with thorium fuels has been in HTRs (see information paper on Thorium).

With negative temperature coefficient of reactivity (the fission reaction slows as temperature increases) and passive decay heat removal, the reactors are inherently safe. HTRs therefore are put forward as not requiring any containment building for safety. They are sufficiently small to allow factory fabrication and will usually be installed below ground level.

Three HTR designs in particular—PBMR, GT-MHR, and Areva's SC-HTGR—were contenders for the Next Generation Nuclear Plant (NGNP) project in the USA (see *Next Generation Nuclear Plant* section in the information page on US Nuclear Power Policy). In 2012, Areva's HTR was chosen. However, the only HTR project currently proceeding is the Chinese HTR-PM.

Hybrid Power Technologies have a hybrid-nuclear Small Modular Reactor (SMR) coupled to a fossil-fuel powered gas turbine.

HTR is seeking government support for a prototype "U-Battery" which would run for 5–10 years before requiring refueling or servicing.

1.8 Small Modular Reactors as Renewable Energy Sources

Advanced Small Modular Reactors (SMRs) are a key part of the Department's goal to develop safe, clean, and affordable nuclear power options. The advanced SMRs currently under development in the USA represent a variety of sizes, technology options, and deployment scenarios. These advanced reactors, envisioned to vary in size from a couple megawatts up to hundreds of megawatts, can be used for power

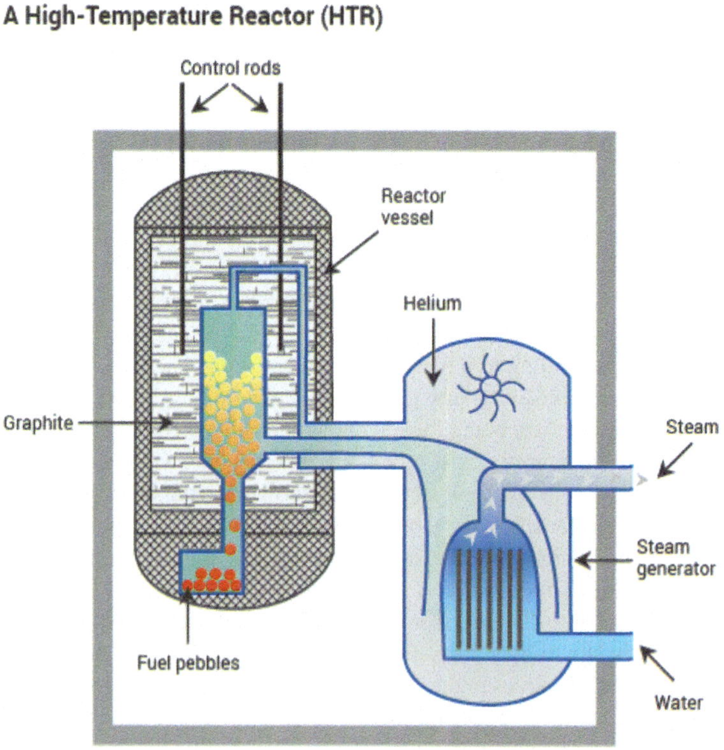

Fig. 1.21 A typical High Temperature Reactor (HTR) configuration

generation, process heat, desalination, or other industrial uses. SMRs can employ light water as a coolant or other non-light water coolants such as a gas, liquid metal, or molten salt [17].

Advanced SMRs offer such as relatively small size, reduced capital investment, ability to be sited in locations not possible for larger nuclear plants, and provisions for incremental power additions. SMRs also offer distinct safeguards, security, and nonproliferation advantages.

The Department of Energy (DOE) has long recognized the transformational value that advanced SMRs can provide to the nation's economic, energy security, and environmental outlook. Accordingly, the Department has provided substantial support to the development of light water-cooled SMRs, which are under licensing review by the Nuclear Regulatory Commission (NRC) and will likely be deployed in the next 10–15 years. The Department is also interested in the development of SMRs that use non-traditional coolants such as liquid metals, salts, and helium because of the safety, operational, and economic benefits they offer.

The US Department of Energy is supporting the development and deployment of advanced SMRs to help meet the nation's economic, environmental, and energy security needs. The Office of Nuclear Energy commissioned three reports in

2017–2018 that examine potential financing opportunities and structures, resiliency credits, and financial incentives for SMRs to help better understand the feasibility of deploying this first-of-a-kind technology in the USA.

Throughout the twentieth century and today, the dramatic increase in energy use for industrial, residential, transportation, and other purposes has been fueled largely by the energy stored in fossil fuels and, more recently, supplied by nuclear power. Many types of renewable electricity generating technologies can be developed and deployed in smaller increments, and constructed more rapidly, than large-scale fossil- or nuclear-based generation systems, thus allowing faster returns on capital investments. However, with new generation of nuclear reactor as we know them as GEN-IV, where small getting smaller to the level of Nuclear Micro Reactor, SMRs are getting more appealing as a new source of renewable energy.

As part of renewable electricity generation technologies, SMRs are under series consideration and they are being looked at. A renewable electricity generation technology harnesses a naturally existing energy flux, such as wind, sun, heat, or tides, and converts that flux to electricity. Natural phenomena have varying time constants, cycles, and energy densities. To tap these sources of energy, renewable electricity generation technologies must be located where the natural energy flux occurs, unlike conventional fossil-fuel and nuclear electricity-generating facilities, which can be located at some distance from their fuel sources.

Renewable technologies also follow a paradigm somewhat different from conventional energy sources in that renewable energy can be thought of as manufactured energy, with the largest proportion of costs, external energy, and material inputs occurring during the manufacturing process. Although conventional sources such as nuclear- and coal-powered electricity generation have a high proportion of capital-to-fuel costs, all renewable technologies, except for biomass-generated electricity (biopower), have no fuel costs. The trade-off is the ongoing and future cost of fossil fuel against the present fixed capital costs of renewable energy technologies.

Scale economics likewise differs for renewables and conventional energy production. Larger coal-fired and nuclear-powered generating facilities exhibit lower average costs of generation than do smaller plants, realizing economies of scale based on the size of the facility. Renewable electricity achieves economies of scale primarily at the equipment manufacturing stage rather than through construction of large facilities at the generating site. Large hydroelectric generating units are an exception and have on-site economies of scale, but not to the same extent as coal- and nuclear-powered electricity plants.

With the exception of hydropower, renewable technologies are often disruptive and do not bring incremental changes to long-established electricity industry sectors. As described by Bowen and Christensen [18], disruptive technologies present a package of performance attributes that, at least at the outset, are not valued by a majority of existing customers. Christensen [19] observes:

Disruptive technologies can result in worse product performance, at least in the near term. Disruptive technologies bring to market very different value propositions than had been available previously. Generally, disruptive technologies underperform established products in mainstream markets. But they have other features that a few fringe customers value. Disruptive technologies that may underperform today, relative to what users in the market demand, may be fully performance-competitive in that same market tomorrow.

Traditional sources of electricity generation at least initially outperform non-hydropower renewables. The environmental attributes of renewables are the initial value proposition that have brought them into the electricity sector. However, with improvements in renewables technologies and increasing costs of generation from conventional sources (particularly as costs of greenhouse gas production are incorporated), renewables may offer the potential to match the performance of traditional generating sources.

1.9 The Limit of Renewable Energy and Small Modular Reactors

Historically, the coal-fired power plants have been generating electricity in the USA and globally to start with and, then with progress in technology of power plant, we turned to fossil fuel and furthermore by improvement of gas turbine, we start using gas fuel power plants to generate electricity for our needs. In 2015, coal plants generated 39% of the 3944 billion kWh of electricity generated in the USA. However, coal's contribution has steadily eroded down from 50% just a decade earlier.

Nuclear power is one of the most recent achievements in the long history of harnessing energy, and one of the most controversial. A result of research originally done to produce the atomic bomb, nuclear energy takes advantage of the incredible potential energy within the atom in a productive instead of destructive way. As of 2011, nuclear energy provides nearly 20% of the electric power in the USA.

Aging infrastructure has made many older and smaller units uneconomical to operate. Nearly 70% of coal-fired generating units comprising more than 50% of the coal generating capacity are more than 40 years old. At the end of 2015, the coal-fired generating units in the USA totaled 286 GW of capacity. In 2015 alone, 11.3 GW of coal-fired capacity were retired. The US Energy Information Administration (EIA) projects that a total of 30 GW of coal-fired generating capacity will retire by 2025, 87% of which by the end of 2020. See Fig. 1.22, where the coal plant shutdown are scheduled.

Tightening environmental regulations have accelerated the trend. New regulatory standards require reductions in emissions of mercury, acid gases, and toxic metals. These standards are scheduled to take effect in April 2015. The capital improvements required to reduce these emissions would make many coal plants uneconomical to operate resulting in the bubble of closures over the next few years. Extensions on compliance are being traded for pledges to close older, dirtier plants completely. The growing public concern with climate change and CO_2 emissions further increases the pressure for coal plant closures.

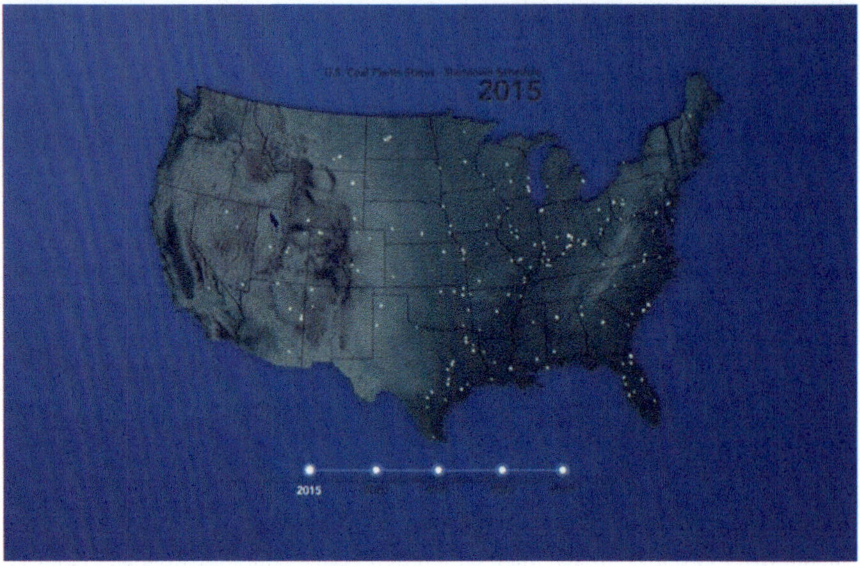

Fig. 1.22 Coal plant shutdown schedule

When it comes to replacing retiring generation, coal now faces greater competition from another fossil fuel, natural gas. Due to technological advances, accessible natural gas reserves have increased dramatically. Natural gas is now available in greater quantities and at low prices. While cleaner and, for the moment, cheaper, natural gas still produces substantial CO_2 emissions and fuel prices are volatile.

Along with the circumstance above, what comes to play was Nuclear Energy that stated for peace around 1950' time frame, where it went through so many generations (see this chapter), and now Generation-III of these power plant is opening room for new and advanced Generation-IV.

If we compare each source of energy to nuclear one as illustrated in Fig. 1.23, we obviously can see that nuclear goes a long way.

However, one may argue if we consider nuclear power energy as a big revolution and evolution in our life since Manhattan project took place, thus the argument may continue on the issue of the bigger is not necessarily the better energy resource.

However, it is very clear that nuclear energy can play a very significant long-term role for meeting the world's increasing supply and demand for energy, based on growth in population globally, while simultaneously addressing challenges associated with global climate and environmental impact.

In today's need for electricity and new source of renewable energy in a very cost-efficient way has sent many countries/nations of the world, particularly the Asia/Pacific Rim countries into quest of new and innovative source of energy beyond what we have from our past technologies, and they all are actively engaged in a major expansion of their nuclear energy complex. This is to the degree, to which nuclear energy can address a long-term energy solution needs, either globally or

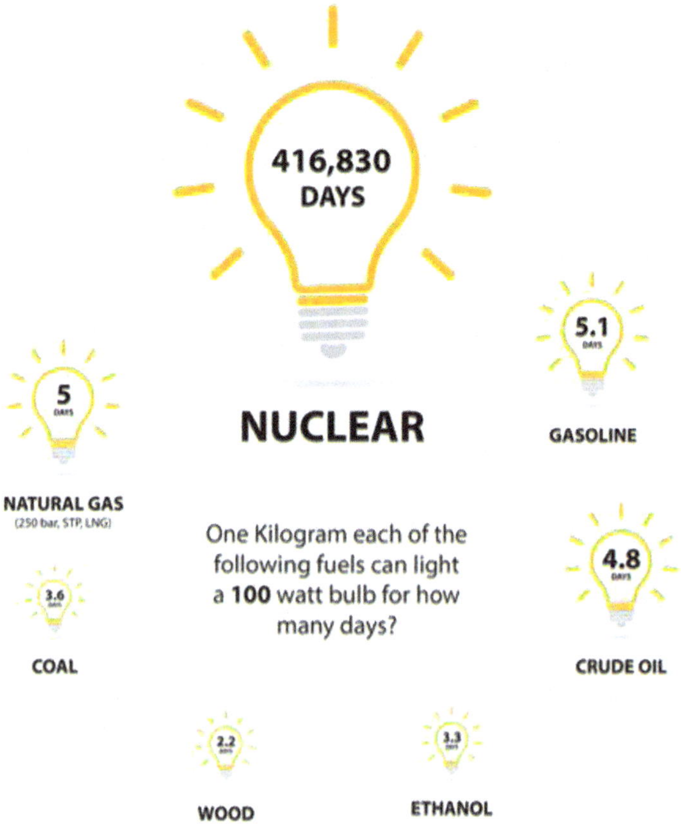

Fig. 1.23 Energy fuel density

regionally, where such resource of energy will be dictated by the peace and adequacy of technical and policy solutions for safety, security, waste management, nonproliferation, and finally greenhouse effects issues, as well as the capital cost of construction, where also energy efficiency has to be challenged. See Fig. 1.24.

Although the chart in Fig. 1.6 is an indication of wind energy to be the most efficient way of producing electrical energy, we have to keep in our mind that not always the wind blowing 24 × 7 around the year and probably there are some regions, that the wind blowing is not as energetic as we needed to be in order to meet the demand for the electricity.

In addition, here, we briefly describe each means of producing electrical energy from each source that is mentioned in Fig. 1.6 as:

1. Biomass
   Everything from crops left in the field to weedy trees, from animal waste to humans' garbage, can be recycled and transformed into usable energy. Biomass is a very broad term covering a wide variety of materials that can be used as

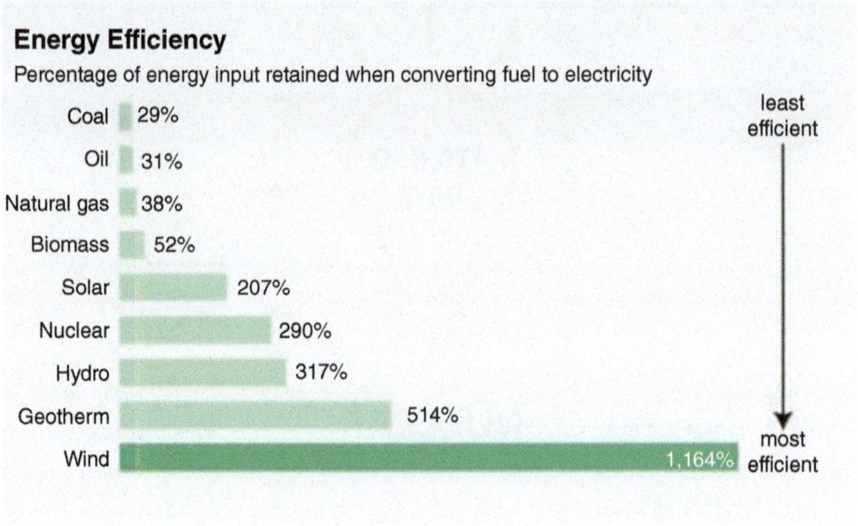

Fig. 1.24 Energy efficiency chart. (Source: Energy Points, The Wall Street Journal)

energy resources. Since the sun's energy is absorbed by all living things, humans, animals, and especially plants, a lot of materials we see as leftovers are storehouses of energy. For example, a tree uses photosynthesis to store energy in its leaves and trunk. The tree is biomass. The tree can be burned to release the energy in the form of heat.

2. Geothermal power
   Geothermal energy is energy that is generated and stored within the earth. It takes advantage of the Earth's core heat to produce usable energy.
3. Hydropower
   Water's raw power provides the energy to produce enough electricity for 28 million Americans every year and, as of 2011, creates nearly 10% all electricity in the USA. Worldwide, hydropower generates more than 2.3 trillion kWh of electricity per year, the energy equivalent to 3.6 billion barrels of oil.
4. Wind power
   One of the most important alternative energy resources cannot be seen or touched, but its power is obvious to anyone who is ever weathered a hurricane, a tornado, or even a strong storm: wind. At its worst, wind can wreak havoc, destroying everything in its path. At its best, it is a source of clean, efficient, inexpensive energy; but as of 2011, it provides less than 3% of all the electricity in the USA.
5. Solar power
   The sun is primarily a source of light and heat. But can it be our primary source of energy? Solar panels or thin films designed to collect sunlight are integral parts of the process to generate electricity by way of the sun. The sun is our most impressive source of energy. More than 1 million times larger than the earth, the

sun gives us ten times more energy than is stored in all the world's reserves of coal and oil every year. Despite this, as of 2011, solar power accounts for less than 1% of all the electricity generated in the USA.

6. Fossil fuels
   Fossil fuels (coal, oil, and natural gas) provide the energy that powers our life styles and our economy. Fossil fuels power everything from the planes in the sky to the cars on the road. They heat our homes and light up the night. They are the bedrock we base our energy mix on. But they are a limited resource.
7. Hydrogen power
   Hydrogen can be used as a way of storing or transporting energy.
8. Energy basics

   Electrical power is produced and distributed through three simple steps:

   (a) generation
   (b) transmission
   (c) distribution

It is also clear that in meeting our low-carbon energy needs nuclear power should play a crucial role. The energy density of nuclear fission means that just a few plants can provide a large percentage of our electricity requirements. In Western liberalized economies, however, traditional large nuclear power plants are not thriving. Struggling utility companies now have difficulty financing projects that can cost upwards of £10 billion and reactor vendors do not have a good record in reducing costs or bringing new plants online on schedule.

Small Modular Reactors (SMRs) could be a solution. Each unit would require a smaller investment than large reactors and their modular nature means that they can be built in a controlled factory environment where, with increased deployment, costs can be brought down over time through improved manufacturing processes and economies of volume. This learning-by-doing effect has helped the offshore wind industry achieve impressive cost reductions and the nuclear industry could replicate their success.

Furthermore, the advanced and innovative SMRs could successfully address several of these issues and offer simpler, standardized, and safer modular design by being factory built, requiring smaller initial capital investment per power plant by virtue of modularity, and having shorter construction time periods. The SMRs, also could be small enough to be portable by means of transportation and occupy smaller real estate due to smaller footprint. It could be implemented in an isolated location without even accessing to the water as coolant media and advanced infrastructure and with no access to power grid (i.e., remote military bases overseas) or could be clustered in a single site to provide a multi-module, large capacity power plant.

To emphasize our argument here in defense of SMRs technology, we can express what Matt Rooney [2] is saying, and that is, "SMRs could offer a number of advantages in a flexible power system, including the potential for dual output, producing other useful services in addition to electricity, like hydrogen or heat. SMRs could, for example, provide a demand/grid management solution by redirecting the power from an SMR to hydrogen production when renewable output is high."

A new fleet of Small Modular Reactors (MSRs) would also provide a large quantity of secure low-carbon energy, thus reducing reliance on imports of natural gas, electricity via interconnectors, and biomass. Uranium, the main source of nuclear fuel, is an inexpensive commodity traded worldwide, and the pioneers in this implementation of source of energy, has the capability to both enrich uranium and manufacture its own nuclear fuel. As we know by now, the nuclear power reduces import dependency from multi-angle point of views.

1.10 Small Modular Reactor-Driven Renewable and Sustainable Energy

In order to address this subject within this section, we too ask ourselves the question; what is the most efficient source of energy?

The answer falls into the following fact that the true cost of electricity is difficult to pin down. That is because a number of inputs comprise: the cost of fuel itself, the cost of production, as well as the cost of dealing with the damage that fuel does to the environment.

Energy Points, a company that does energy analysis for business, factors in these myriad values in terms of what percentage of the energy input—fossil-fuel energy, plus energy for production and energy for environmental mitigation—will become usable electricity.

The chart in Fig. 1.25 shows that fossil fuels yield, on a national average, only a portion of their original energy when converted into electricity. That is because they are fossil fuels that require other fossil fuels to make the conversion into electricity; their emissions, such as carbon dioxide, also require a lot of energy to be mitigated. Renewables, however, have energy sources that are not fossil fuel and their only other energy inputs are production and mitigating the waste from that production. That actually results in more energy produced than fossil fuels put in. Wind, the most efficient fuel for electricity, creates 1164% of its original energy inputs when converted into electricity; on the other end of the efficiency spectrum, coal retains just 29% of its original energy.

These are national averages, meaning that, for example, solar might be more efficient in a place such as Arizona with lots of infrastructure and direct sunlight than it is across the whole nation. Thus, a scenario such as Solar Farm technology may very well be suited in such environment and arguably source of fresh water shortage as a coolant media for fossil, gas and nuclear power plant, may also enhance the solar power plant farm as only choice of solution to generate electricity as well renewable energy approach, which may very well be cost-efficient for such production.

However, no matter what, in any given area, electricity might come from a number of different sources, including oil, coal, gas, wind, hydropower, and solar. Each has its own set of costs, both internal and external. From Energy Points:

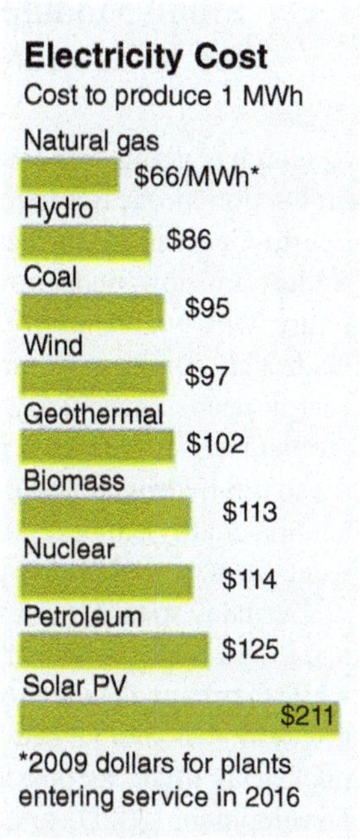

Fig. 1.25 Cost of electricity per 1 MWh. (Source: EIA, The Wall Street Journal)

Energy Points' methodology measures environmental externalities and calculates the energy it takes to mitigate them. For example, it quantifies the Green-House Gas (GHG) emissions that result from turning coal and natural gas into electricity and then calculates the energy it would take to mitigate those emissions through carbon capture and sequestration. Water scarcity and contamination are quantified as the energy that is required to durably supply water to that area. And in the case of solar or wind energy, Energy Points incorporates the life cycle impact of manufacturing and shipping the panels.

This metric is a more rounded calculation than merely cost or carbon footprint. For example, hydroelectricity has the lowest carbon footprint of 4 gCO_2/kWh, but when Energy Points factors in the full life cycle of the different fuels, wind is the most efficient. Additionally, natural gas is the cheapest fuel to produce electricity, according to levelized cost data from the Environmental Protection Agency, which measured the total cost of building and operating a generating plant over an assumed financial life and duty cycle. Though it is cheap, it is not very efficient if you factor in its production and emissions.

1.11 Small Modular Reactor-Driven Hydrogen Energy for Renewable Energy Source

Research is going forward to produce hydrogen based on nuclear energy. Hydrogen production processes necessitate high temperatures that can be reached in the Fourth-Generation nuclear reactors (i.e., Small Modular Reactors). Technological studies are now underway in order to define and qualify components that in the future will enable us to retrieve and transfer heat produced by these reactors. Hydrogen combustion turbine power could be one of the solutions to our future energy needs particularly in on-peak demand for electricity, but until recently the problem with hydrogen power was its production for use as an energy source. Although hydrogen is the most common element in the known universe to human being, actually capturing it for energy use is a process which itself usually requires some form of fuel or energy [20].

Germany to take a drastic measurement to revise their nuclear energy policy that had long heralded nuclear power plants as its main source of energy. For example, while Germany decided to abandon all of their atomic power plants, the new energy policy that is announced by Japan is taking steps to decrease its dependency on nuclear as much as possible, while increasing and enhancing their Research and Development (R&D) to quest for an alternative renewable energy source. Also, in parallel effort the governed is promoting for a "Hydrogen Society" and use of hydrogen as source of energy to pave their way to such goal, by making, for example, Fuel Cell Vehicle (FCV), where Fuel Cell and Hydrogen Technology Group at the New Energy and Industrial Technology Development Organization (NEDO) is charge of such R&D [17].

Burning hydrogen in a combustion form that does not emit any carbon dioxide (CO_2), so it is considered as a source of clean energy that can greatly help reduce the greenhouse gases effects. Although expectations are set so high, there comes with the technical challenges and cost of ownership as well as return on investment of such research and development toward full production such source of energy as part of renewable form. As an example, setting up expensive hydrogen stations for FCVs, securing sufficient supplies of the gas and coming up with ways to produce it without emitting carbon dioxide are just a few of those challenges and hurdles [20].

Other industrial application of hydrogen is in oil refinery, where it is used to process crude oil into refined fuel, such as gasoline and diesel, and for removing contaminants, such as sulfur, from these fuels. See Fig. 1.26.

Total hydrogen consumption in oil refineries is estimated at 12.4 billion standard cubic feet per day, which equates to an average hydrogen consumption of 100–200 standard cubic feet per barrel of oil processed. Hydrogen consumption in the oil refining industry grew at a compound annual growth rate of 4% from 2000 to 2003, and growth in consumption is expected to increase between 5% and 10% through to 2010 [Oil & Gas Journal, CryoGas International]. See Fig. 1.27 [21].

Fig. 1.26 A typical oil refinery plant

The principal drivers of this growth in refinery hydrogen demand are:

- Low sulfur in diesel fuel regulations—hydrogen is used in refineries to remove sulfur from fuels such as diesel.
- Increased consumption of low-quality "heavy" crude oil, which requires more hydrogen to refine.
- Increased oil consumption in developing economies such as China and India.

Approximately 75% of hydrogen currently consumed worldwide by oil refineries is supplied by large hydrogen plants that generate hydrogen from natural gas or other hydrocarbon fuels, with the balance being recovered from hydrogen-containing streams generated in the refinery process.

Pressure Swing Adsorption (PSA) (see Figs. 1.28 and 1.29) technology is used in both hydrogen generation plants and for hydrogen recovery.

Hydrogen is used in a range of other industries, including chemical production, metal refining, food processing, and electronics manufacturing. Hydrogen is either delivered to customers in these industries as compressed or liquid hydrogen or generated on-site from water using a process known as electrolysis or from natural gas using a process call reforming. In certain applications, there is a gradual shift toward on-site generation to replace delivered compressed or liquid hydrogen, largely based on the lower cost of new on-site hydrogen generation technologies when compared to delivered hydrogen [20].

Fig. 1.27 Hydrogen PSA unit—HYDROSWING. (Courtesy of Mahler Advanced Gas Systems)

Fig. 1.28 A typical metal refining plant

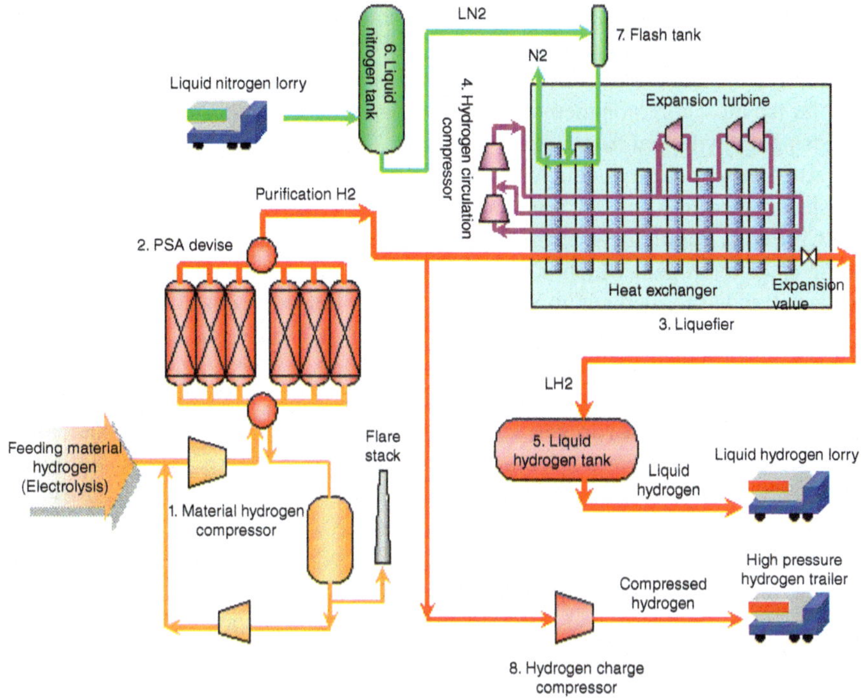

Fig. 1.29 Flow of liquid hydrogen production facilities

Other applications of hydrogen in industry that worth to mention are listed as below:

1. Weather balloons in meteorologist, where these balloons are fitted with equipment to record information necessary to study the climates.
2. Hydrogen is used in fertilizer and paint industries.
3. Food industries, where in food it is used as element to make hydrogenated vegetable oils, while using nickel as a catalyst, solid fat substances are produced.
4. Welding companies use the hydrogen as part of welding torches element. These torches are utilized for steel melting.
5. Chemical industries use them for metal extraction. For example, hydrogen is needed to treat mined tungsten to make them pure.
6. In home uses, hydrogen peroxide can be used in non-medical ways. Other applications include a pest controller in gardens, removing stains on clothing and functioning as a bleaching agent for cleaning homes.

As we can see, hydrogen is an important utility for numerous applications in multiple industries. Users in a wide range of industries can benefit from operating a cost-effective hydrogen plant and reduce their production costs significantly. See Fig. 1.9.

As we know from our knowledge of chemistry, hydrogen is the lightest and most common element in the cosmos. Its atomic number is 1. In its elemental state, hydrogen is rare. But it is one of the components of water and vital to life. Hydrogen alone does not exist as a natural resource, and it needs to be produced by separating from other elements and molecules, such water as we have vast oceans sounding us. By far the most common method of producing hydrogen in industry currently is due to stripping hydrogen from natural gas using a process known as steam reforming.

Another way of producing hydrogen is through electric hydrolysis as an alternate to steam reforming approach and both methods were mentioned as below.

Currently, fossil fuels, including naphtha, natural gas and coal, are the main sources of hydrogen, which is generated by "steam reforming" method, in which steam is added to methane to yield hydrogen. A huge amount of hydrogen is also produced as a by-product from the production of caustic soda plants and from coke ovens.

In contrast, electric hydrolysis is a relatively simple process and methods in which production of hydrogen takes place any high school chemistry laboratory course, where two electrodes, one with positive charge known as anode and other negatively charged know as cathode by a battery are placed into water. Result of such induced electric current through water splits the hydrogen ion from oxygen with positive hydrogen ion being attracted to cathode a negative oxygen ion goes toward anode. Once the ions touch the electrodes the hydrogen gains and electron while oxygen loses one and they are creating fully fledged atoms of hydrogen and oxygen, which then rise in the water and they can be collected separately at the top of water container.

The Japanese organization, NEDO published a white paper on hydrogen energy in July 2012 that states the importance of promoting hydrogen-related products, which in Japan are expected to develop into a market worth ¥1 trillion by 2030 and ¥8 trillion by 2050 [22].

References

1. B. Zohuri, P. McDaniel, *Advanced Smaller Modular Reactors: An Innovative Approach to Nuclear Power*, 1st edn. (Springer, Cham, 2019). https://www.springer.com/us/book/9783030236816
2. B. Zohuri, P. McDaniel, *Combined Cycle Driven Efficiency for Next Generation Nuclear Power Plants: An Innovative Design Approach*, 2nd edn. (Springer, Cham, 2018). https://www.springer.com/gp/book/9783319705507
3. B. Zohuri, P. McDaniel, C.R. De Oliveria, Advanced nuclear open air-Brayton cycles for highly efficient power conversion. Nucl. Technol. **192**(1), 48–60 (2015). https://doi.org/10.13182/NT14-42
4. B. Zohuri, *Combined Cycle Driven Efficiency for Next Generation Nuclear Power Plants: An Innovative Design Approach* (2016)
5. https://en.wikipedia.org/wiki/Nuclear_and_radiation_accidents_and_incidents
6. https://www.cbc.ca/news/technology/nuclear-capacity-climate-goals-power-supply-iea-1.5152080
7. https://www.ft.com/content/bcffe4d2-2402-11e6-9d4d-c11776a5124d

8. B. Zohuri, *Magnetic Confinement Fusion Driven Thermonuclear Energy*, 1st edn. (Springer Publishing Company, Cham, 2017)
9. B. Zohuri, *Inertial Confinement Fusion Driven Thermonuclear Energy*, 1st edn. (Springer Publishing Company, Cham, 2017)
10. https://www.world-nuclear.org/information-library/nuclear-fuel-cycle/nuclear-power-reactors/small-nuclear-power-reactors.aspx
11. https://en.wikipedia.org/wiki/MOX_fuel
12. B. Zohuri, *Heat Pipe Application in Fission Driven Nuclear Power Plants*, 1st edn. (Springer Publishing Company, Cham, 2019)
13. B. Zohuri, *Heat Pipe Design and Technology: Modern Applications for Practical Thermal Management*, 2nd edn. (Springer Publishing Company, Cham, 2016)
14. K. Stacey, Small Modular Reactors Are Nuclear Energy's Future, *Financial Times*, 25 July 2016
15. Small Modular Reactors: A Window on Nuclear Energy, Andlinger Center, Princeton University, June 2015 http://acee.princeton.edu/distillates
16. https://theconversation.com/the-nuclear-industry-is-making-a-big-bet-on-small-power-plants-94795
17. B. Zohuri, *Small Modular Reactors as Renewable Energy Sources*, 1st edn. (Springer Publishing Company, Cham, 2019)
18. Bowen and Christensen, "*Disruptive Technologies: Catching the Wave*" HBR, January-February (1995)
19. CM. Christensen, *Dilemma When New Technologies Cause Great Firms to Fail*, Harvard Business Review Press, Boston, Massachustts (1997)
20. B. Zohuri, *Nuclear Energy for Hydrogen Generation through Intermediate Heat Exchangers: A Renewable Source of Energy* (Springer Publishing Company, Cham, 2016)
21. http://www.xebecinc.com/applications-industrial-hydrogen.php
22. http://www.japantimes.co.jp/news/2014/10/12/national/japan-rises-challenge-becoming-hydrogen-society

Chapter 2
Nuclear Industry Trend Toward Small and Micro Nuclear Power Plants

2.1 Introduction

Today, due partly to the high capital cost of large power reactors generating electricity via the steam cycle and partly to the need to service small electricity grids under about 4 GWe, there is a move to develop smaller units. These may be built independently or as modules in a larger complex, with capacity added incrementally as required (see Sect. 2.2). Economies of scale are envisaged due to the numbers produced. There are also moves to develop independent small units for remote sites. Small units are seen as a much more manageable investment than big ones whose cost often rivals the capitalization of the utilities concerned.

An additional reason for interest in SMRs is that they can more readily slot into brownfield sites in place of decommissioned coal-fired plants, the units of which are seldom very large—more than 90% are under 500 MWe, and some are under 50 MWe. In the USA, coal-fired units retired over 2010–12 averaged 97 MWe, and those expected to retire over 2015–25 average 145 MWe [1].

Small Modular Reactors (SMRs) are defined as nuclear reactors generally 300 MWe equivalent or less, designed with modular technology using module factory fabrication, pursuing economies of series production and short construction times. This definition, from the World Nuclear Association, is closely based on those from the International Atomic Energy Agency (IAEA) and the US Nuclear Energy Institute. Some of the already-operating small reactors mentioned or tabulated below do not fit this definition, but most of those described do fit it [1].

SMR development is proceeding in Western countries with a lot of private investment, including small companies. The involvement of these new investors indicates a profound shift taking place from government-led and -funded nuclear R&D to that led by the private sector and people with strong entrepreneurial goals, often linked to a social purpose. That purpose is often deployment of affordable clean energy, without carbon dioxide emissions.

B. Zohuri, *Nuclear Micro Reactors*, https://doi.org/10.1007/978-3-030-47225-2_2

As we stated in the preface of the book, nuclear reactor is getting smaller and it is opening up some big opportunities for the industry that are involved with design and manufacturing of these reactors.

A handful of micro reactor designs are under development in the USA, and they could be ready to roll out within the next decade.

These plug-and-play reactors will be small enough to transport by truck and could help solve energy challenges in a number of areas, ranging from remote commercial or residential locations to military bases. As illustrated in Fig. 2.1, micro reactors have a big potentials.

Some of their features that we can state that Nuclear Micro Reactors (NMRs) are not defined by their fuel form or coolant. Instead, they have three main features:

1. *Factory fabricated*: All components of a micro reactor would be fully assembled in a factory and shipped out to location. This eliminates difficulties associated with large-scale construction, reduces capital costs, and would help get the reactor up and running quickly.
2. *Transportable*: Smaller unit designs will make micro reactors very transportable. This would make it easy for vendors to ship the entire reactor by truck, shipping vessel, airplane, or railcar.
3. *Self-regulating*: Simple and responsive design concepts will allow micro reactors to self-regulate. They will not require a large number of specialized operators and would utilize passive safety systems that prevent any potential for overheating or reactor meltdown.

The benefits of these type of reactors lays in their design. Micro reactor designs vary, but most would be able to produce 1–20 MW of thermal energy that could be used directly as heat or converted to electric power. They can be used to generate clean and reliable electricity for commercial use or for non-electric applications such as district heating, water desalination, and hydrogen fuel production.

Other benefits are included as follows:

- Seamless integration with renewables within microgrids.
- Can be used for emergency response to help restore power to areas hit by natural disasters.
- A longer core life, operating for up to 10 years without refueling.
- Can be quickly removed from sites and exchanged for new ones.

Most designs will require fuel with a higher concentration of uranium-235 that is not currently used in today's reactors although some may benefit from use of high temperature moderating materials that would reduce fuel enrichment requirements while maintaining the small system size.

The US Department of Energy supports a variety of advanced reactor designs, including gas, liquid metal, molten salt, and heat pipe-cooled concepts. American micro reactor developers are currently focused on gas and heat pipe-cooled designs that could debut as early as the mid-2020s.

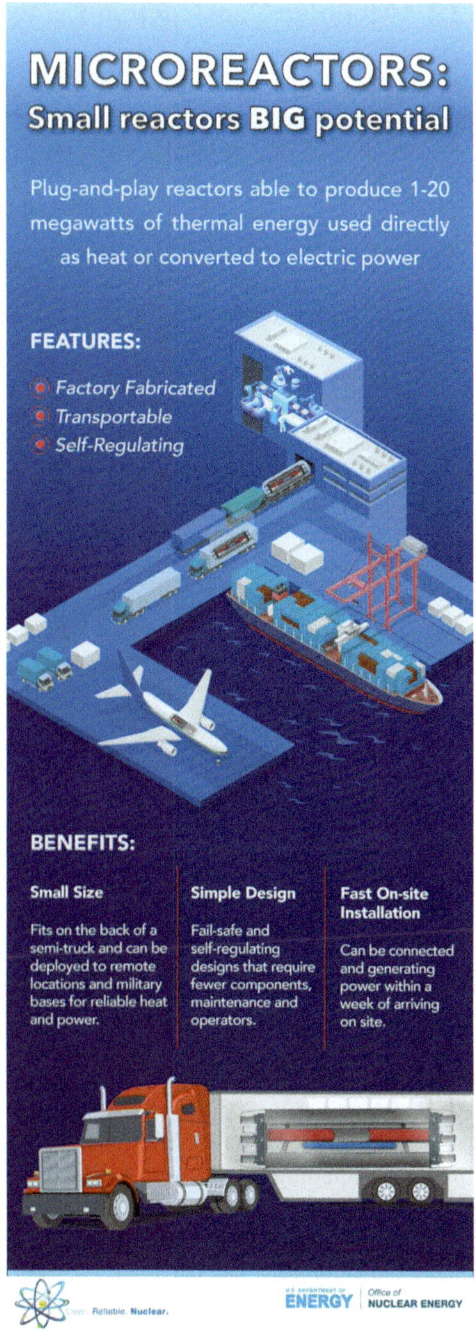

Fig. 2.1 Big potential of small reactors. (Source: US Energy Information Administration)

2.2 Modular Construction Using Small Reactor Units

Westinghouse and IRIS partners have outlined the economic case for modular construction of their IRIS design (about 330 MWe), and the argument applies similarly to other similar or smaller units. They pointed out that IRIS with its size and simple design is ideally suited for modular construction in the sense of progressively building a large power plant with multiple small operating units. The economy of scale is replaced here with the economy of serial production of many small and simple components and prefabricated sections. They expected that construction of the first IRIS unit would be completed in 3 years, with subsequent reduction to only 2 years.

Site layouts have been developed with multiple single units or multiple twin units. In each case, units will be constructed so that there is physical separation sufficient to allow construction of the next unit while the previous one is operating and generating revenue. In spite of this separation, the plant footprint can be very compact so that a site with, for instance, three IRIS single modules providing 1000 MWe capacity would be similar or smaller in size than one with a comparable total power single unit.

Many small reactors are designed with a view to serial construction and collective operation as modules of a large plant. In this sense, they are "small modular reactors"—SMRs—but not all small reactors are of this kind (e.g., the Toshiba 4S) though the term SMR tends to be used loosely for all small designs.

Eventually plants comprising a number of SMRs are expected to have a capital cost and production cost comparable with larger plants. But any small unit such as this will potentially have a funding profile and flexibility otherwise impossible with larger plants. As one module is finished and starts producing electricity, it will generate positive cash flow for the next module to be built. Westinghouse estimated that 1000 MWe delivered by three IRIS units built at 3 year intervals financed at 10% for 10 years require a maximum negative cash flow less than $700 million (compared with about three times that for a single 1000 MWe unit). For developed countries, small modular units offer the opportunity of building as necessary; for developing countries, it may be the only option because their electric grids cannot take 1000+ MWe single units. The Westinghouse SMR is a >225 MWe integral Pressurized Water Reactor (PWR) with all primary components located inside the reactor vessel.

The Westinghouse Small Modular Reactor (SMR) as illustrated in Fig. 2.2, is an 800 MWt/225 MWe class integral PWR with passive safety systems and reactor internals including fuel assemblies based closely on those in the AP1000 (89 assemblies 2.44 m active length, <5% enrichment).

The steam generator is above the core fed by eight horizontally mounted axial-flow coolant pumps. The reactor vessel will be factory-made and shipped to site by rail, then installed below ground level in a containment vessel 9.8 m diameter and 27 m high. The reactor vessel module is 25 m high and 3.5 m diameter. It has a 24-month refueling cycle and 60-year service life. Passive safety means no operator intervention is required for 7 days in the event of an accident. Daily load following

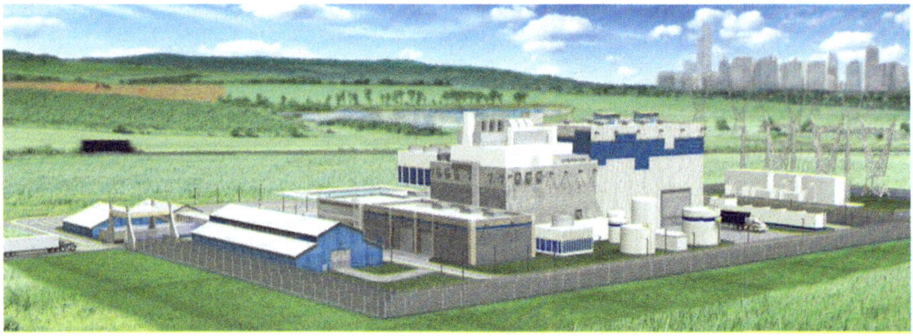

Fig. 2.2 Artistic depiction of Westinghouse Small Modular Reactor

can be performed from 100% to 20% power at a rate of 5% change per minute; in continuous load following, the plant can perform load changes of ±10% power at a rate of 2% per minute.

Westinghouse SMR utilizes passive safety systems and proven components—realized in the industry-leading AP1000® reactor design—to achieve the highest level of safety and reduce the number of components required. This approach will provide licensing, construction, and operation certainty that no other SMR supplier can match.

The Westinghouse Small Modular Reactor offers the following advantages and are included as:

- Passive safety features designed to shut the plant down automatically and keep it cool without human intervention or AC power for 7 days.
- Reduced fuel, resulting in reduced radioactivity amounts released in the case of an accident.
- Passive heat removal with on-site water inventory, which relies on the natural forces of evaporation, condensation, and gravity.
- Underground containment.
- Innovative, integral design eliminates a number of accident scenarios.
- Improved energy security and reduction in the overall life cycle carbon footprint when used to provide power for liquid transportation fuel from resources of oil sands, oil shale, and coal-to-liquid applications.

Comparison of other clean energy source with the Westinghouse SMR is illustrated in Fig. 2.3 here.

Innovative load of this SMR following capabilities are listed as:

- The SMR is capable of economically handling the unique challenges of providing baseload power on smaller grids and those with non-steady power sources.
- The SMR utilizes the Westinghouse mechanical shim (MSHIM™) operating strategy to follow grid changes.
- MSHIM allows easy transition between load follow and baseload operation with minimal operator interaction.

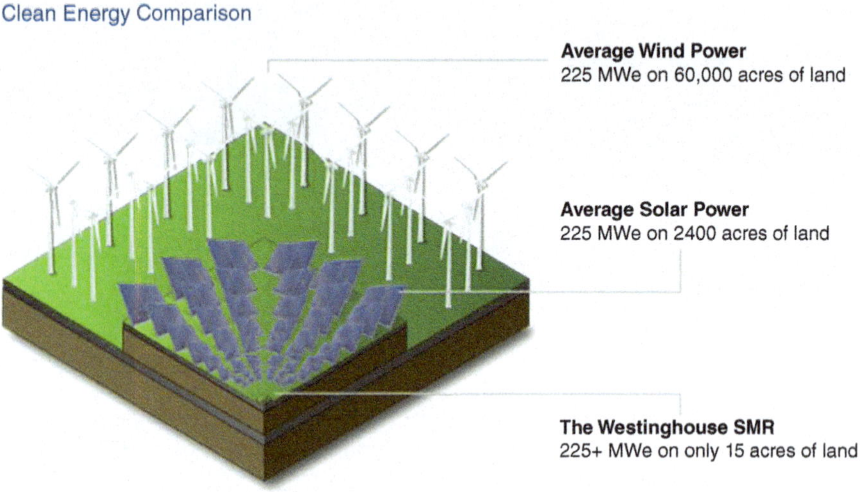

Fig. 2.3 Clean energy comparison

- Operational costs are lowered due to minimized chemistry change requirements.
- Daily load follow can be performed from 100% to 20% power at a rate of 5% change per minute; in continuous load follow, the plant can perform load changes of ±10% power at a rate of 2% per minute.

Finally, the Westinghouse Small Modular Reactor features as illustrated in Fig. 2.4 at a glance are listed below:

- Electric output: >225+ MWe
- Reactor power: 800 MWt
- Design life: 60 years
- Fuel type: 17 × 17 RFA, <5% enriched UO_2
- Total site area: ~15 acres
- Passive safety systems
- Rail, truck, or barge shippable
- Compact integral design
- Simplified system configuration, standardized, fully modular approach
- Minimized footprint, maximized power output
- 24 Months between refueling

As part of nuclear industry efforts to make small construction of these reactors smaller, Westinghouse approach toward Nuclear Micro Reactor is appeared in the form of eVinci ™ that Westinghouse is currently developing this reactor.

eVinci micro reactor, a next-generation, is very small modular reactor for decentralized generation markets.

The eVinci micro reactor's innovative design is a combination of nuclear fission, space reactor technologies, and 50+ years of commercial nuclear systems design, engineering, and innovation. The eVinci micro reactor aims to create affordable and

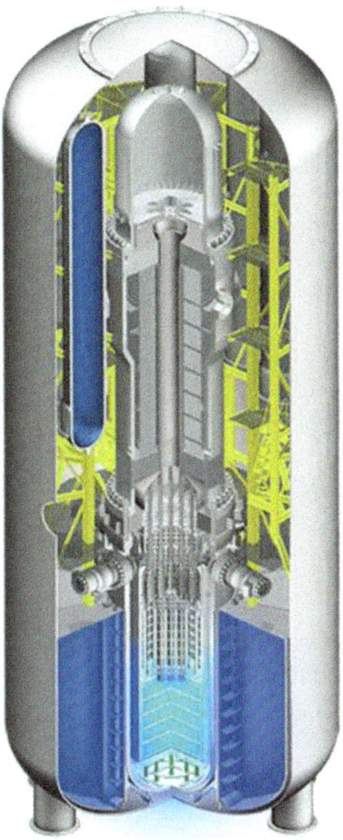

Fig. 2.4 The Westinghouse SMR Core

sustainable power with improved reliability and minimal maintenance, particularly for energy consumers in remote locations. The small size of the generator allows for easier transportation and rapid, on-site installation in contrast to large, centralized stations. The reactor core is designed to run for more than 10 years, eliminating the need for frequent refueling.

The key benefits of the eVinci micro reactor are attributed to its solid core and advanced heat pipes technologies [2, 3]. The core encapsulates fuel to significantly reduce proliferation risk and enhances overall safety for the user. The heat pipes enable passive core heat extraction and inherent power regulation, allowing autonomous operation and inherent load following capabilities. These advanced technologies together make the eVinci micro reactor a pseudo "solid-state" reactor with minimal moving parts and as illustrated in Fig. 2.5 is very mobile and portable as well.

As part of our question that was stated in Sect. 1.4 "How Small is Small?", we may say the small nuclear reactors are not anything actually new technologically. India has the most, with 18 reactors with capacity ranging between 90 and 220 MW, which were built between 1981 and 2011.

The United States, Russia, China, India, and France with the smallest nuclear submarines known as *Rubis-Class* attack boat (i.e., Fig. 2.6) with 2600 dwt and the UK with *Vanguard Class* Ballistic missile submarine (SSBN) of 15,900 dwt

Fig. 2.5 eVinci micro reactor held in a transport container. (Source: Courtesy of Westinghouse Corporation)

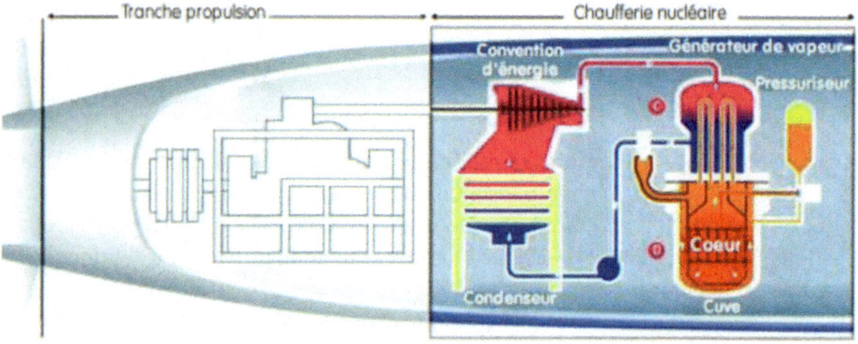

Fig. 2.6 French integral Pressurized Water Reactor (PWR) for submarine

submerged that has a single reactor (i.e., Fig. 2.7) operate hundreds of nuclear submarines and aircraft carriers. Russia has dozens of nuclear-powered icebreakers cruising around the Arctic, and its first floating nuclear power plant has been completed and will be deployed in 2019 near the town of Pevek in East Siberia.

The Siberian plant will replace four 12-MW reactors the Soviets built in the 1970s to power a remote town and administrative center, as well as mining and oil drilling operation.

Bear in your mind that the work on nuclear power ships and nuclear marine propulsion first was started in the 1940s, and the first test reactor started up in the USA in 1953. The first nuclear-powered submarine, *USS Nautilus*, put to sea in 1955 (i.e., Fig. 2.8).

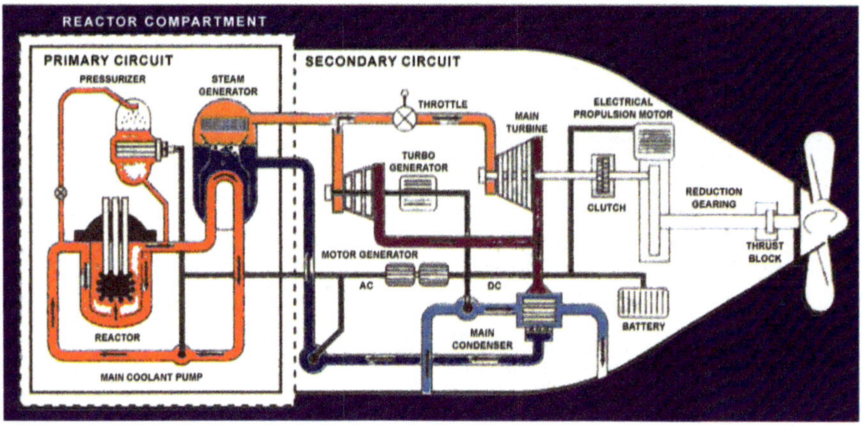

Fig. 2.7 UK Nuclear submarine layout

Fig. 2.8 Portrait of USS Nautilus Submarine

This marked the transition of submarines from slow underwater vessels to warships capable of sustaining 20–25 knots submerged for weeks on end. The submarine had come into its own.

Nautilus led to the parallel development of further (*Skate*-class) submarines, powered by single Pressurized Water Reactors (PWRs), and an aircraft carrier, *USS*

Enterprise (i.e., Fig. 2.9), powered by eight Westinghouse reactor units in 1960. A cruiser, *USS Long Beach* (i.e., Fig. 2.10), followed in 1961 and was powered by two of these early units. Remarkably, the *Enterprise* remained in service to the end of 2012.

Other countries such as China also involved with small reactors. China General Nuclear Group (CGN) has two small ACPR designs: an ACPR100 and ACPR50S, both with passive cooling for decay heat and 60-year design life. Both have standard type fuel assemblies and fuel enriched to <5% with burnable poison giving 30-month refueling. The ACPR100 is an integral PWR, 450 MWt, 140 MWe, having 69 fuel assemblies. Reactor pressure vessel is 17 m high and 4.4 m inside diameter, operating at 310 °C. It is designed as a module in larger plant and would be installed underground. The applications for these are similar to those for the ACP100.

As illustrated in Fig. 2.11, the Ocean Star-V version would be on a barge, as a floating nuclear power plant.

The offshore ACPR50S is 200 MWt, 60 MWe with 37 fuel assemblies, and four external steam generators. Reactor pressure vessel is 7.4 m high and 2.5 m inside diameter, operating at 310 °C. It is designed for mounting on a barge as Floating Nuclear Power Plant (FNPP). Following approval as part of the 13th Five-Year Plan for innovative energy technologies, CGN announced construction start on the first FNPP at Bohai shipyard in November 2016 for trial operation in 2019, supplying power and desalination.

Fig. 2.9 USS Enterprise (CVN-80)

Fig. 2.10 USS Long Beach Cruiser

Fig. 2.11 CGN's floating reactor concept

Following our discussion how small is small, we may say that, even though the reactors will be small, they may operate at much bigger power plants with multiple reactors. NuScale, for example, wants to install 12 reactors at its initial Idaho site. Based on the company's latest projections, it will have a total capacity of 720 MW.

A global trend toward small and smaller reactors are gained momentum in recent years starting with Generation IV (GEN-IV) reactors research study and designs as well as fabrication of them. Private and state-owned companies are seeking to build these small power plants in about a dozen countries so far, including the USA and the UK.

France, which gets three-quarters of its electricity from nuclear energy, and Canada may soon join the fray.

This global interest in small modular reactors comes as more standard nuclear reactors are being decommissioned than are under construction.

Pushing the Small Modular Reactors (SMRs) to smaller footnote is led by the Westinghouse Corporation one of pointers in Nuclear Power Industry by developing eVinci micro reactor. The eVinci micro reactor's innovative design is a combination of nuclear fission, space reactor technologies, and 50+ years of commercial nuclear systems design, engineering, and innovation. The eVinci micro reactor aims to create affordable and sustainable power with improved reliability and minimal maintenance, particularly for energy consumers in remote locations. The small size of the generator allows for easier transportation and rapid, on-site installation in contrast to large, centralized stations. The reactor core is designed to run for more than 10 years, eliminating the need for frequent refueling.

The key benefits of the eVinci micro reactor are attributed to its solid core and advanced heat pipes. The core encapsulates fuel to significantly reduce proliferation risk and enhances overall safety for the user. The heat pipes enable passive core heat extraction and inherent power regulation, allowing autonomous operation and inherent load following capabilities. These advanced technologies together make the eVinci micro reactor a pseudo "solid-state" reactor with minimal moving parts.

Key Attributes of eVinci Micro Reactor:

- Transportable energy generator.
- Fully factory built, fueled, and assembled.
- Combined heat and power—200 kWe to 5 MWe.
- Up to 600 °C process heat.
- 5–10-year life with walkaway inherent safety.
- Target less than 30 days on-site installation.
- Autonomous load management capability.
- Unparalleled proliferation resistance.
- High reliability and minimal moving parts.
- Green field decommissioning and remediation.

Figure 2.12 pretty much is indication of these attributes.

The Department of Energy (DOE) is funding a venture that will put together Westinghouse's 25-MWe eVinci micro reactor for nuclear demonstration readiness by 2022.

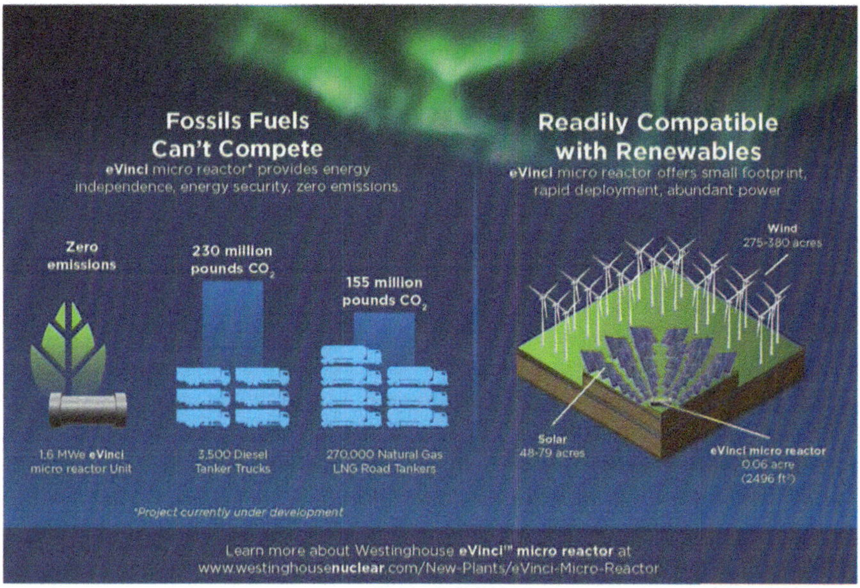

Fig. 2.12 Westinghouse eVinci™ micro reactor attributions

The company on March 27 stated it is going to present \$12.9 million of the estimated \$28.6 million Westinghouse wants for a venture to arrange the micro reactor for an illustration, together with for design, evaluation, licensing to fabricate, siting, and testing. eVinci is one in all three Small Modular Reactors (SMRs) and the primary micro reactor whose first-of-a-kind growth the DOE is subsidizing below a December 2017–issued "U.S. Industry Opportunities for Advanced Nuclear Technology Development" funding alternative announcement (FOA). The Funding Opportunity Announcement (FOA) advertises availability of grant support.

The funding bulletins are a part of a latest ramp up in consideration and efforts by the US authorities to spice up growth of superior nuclear applied sciences. Also on March 27, a bipartisan group of senators launched laws to bolster innovation for superior reactors. Lawmakers stated that the USA, which as soon as led efforts to invent and commercialize key nuclear applied sciences, has "slipped" in management, and it dangers dropping out to Russia and China.

2.3 A Novel Heat Pipe Reactor

For Westinghouse, the federal funding poses a major opportunity to boost development of its novel micro reactor technology in a highly competitive environment. (For more details, see "Big Gains for Tiny Nuclear Reactors," in POWER's November 2018 issue) [4]. According to Westinghouse, the eVinci reactor is an innovative combination of nuclear fission and space reactor technologies that integrates the company's long-standing experience in commercial nuclear systems design, engineering, and innovation (i.e., Fig. 2.13).

The small size of the generator allows for easier transportation and rapid, on-site installation in contrast to large, centralized stations, the company says (i.e., Fig. 2.14). And because the reactor core is designed to run for more than 10 years, eliminating the need for frequent refueling, Westinghouse is marketing it as an off-grid or microgrid solution.

But, according to some experts, eVinci's technological simplicity is what makes it unique. The reactor is envisioned to operate autonomously. Its reactor core is a solid-steel monolith that features channels for fuel pellets, the moderator (metal hydride), and heat pipes, which are arranged in a hexagonal pattern. See Fig. 2.15.

Figure 2.15 is showing the core and surrounding structure cross-section of eVinci Novel Heat Pipe Micro Reactor, while Fig. 2.16 is layout of same reactor presenting the passive cooling system of heat pipes in it.

The monolith will serve as the second fission product barrier (the fuel pellet is the first barrier) as well as the thermal medium between the fuel channels and heat pipes. The heat pipes will extract heat from the core using a technology based on thermal conductivity and fluid phase transition.

However, Westinghouse admits that it is fielding a number of challenges related to the deployment of the micro reactor. While the eVinci will use fuel enriched to 19.75 wt%, the industrial scale amount of uranium enriched to more than 5% is limited. The DOE is scrambling to resolve that issue, announcing in January that it would demonstrate—by October 2020—the production of high-assay low-enriched uranium (HALEU) fuel at the Portsmouth Gaseous Diffusion Plant in Piketon, Kentucky. On February 27, meanwhile, Maryland-based X-Energy, another DOE funding recipient (see below), dedicated its HALEU-based TRISO-X fuel fabrication pilot line at the Oak Ridge National Laboratory.

Westinghouse also notes that while eVinci reactors will be manufactured and assembled in a factory, first reactor startup should also happen at that site, which means the factory will need to be equipped with radio protection equipment, safety and security systems, and have a license from the Nuclear Regulatory Commission (NRC).

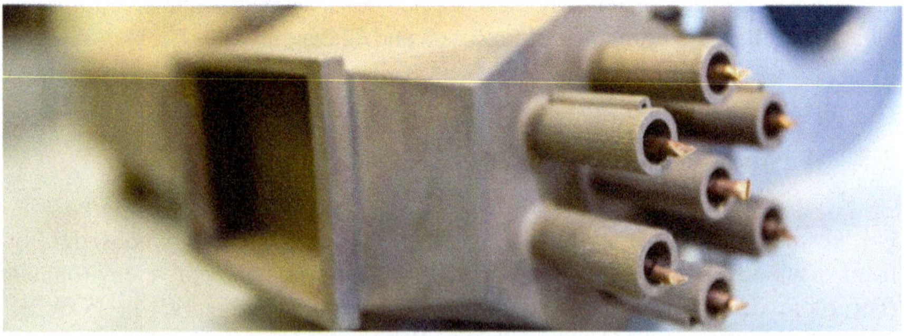

Fig. 2.13 eVinci™ micro reactor augmenting heat pipe

Fig. 2.14 Transportability of eVinci illustration

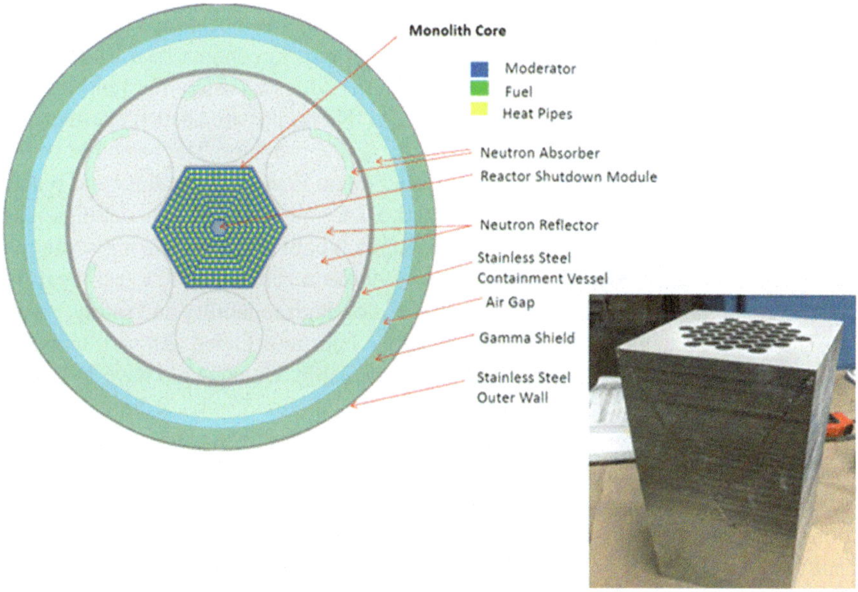

Fig. 2.15 A cross-section of Westinghouse's eVinci reactor. (Source: Courtesy of Westinghouse Corporation)

Transportation must also take into account safety and security. And because the reactor will be operated autonomously, Westinghouse will need to field first-of-their-kind challenges in licensing, instrumentation, remote reactor monitoring, and logistics.

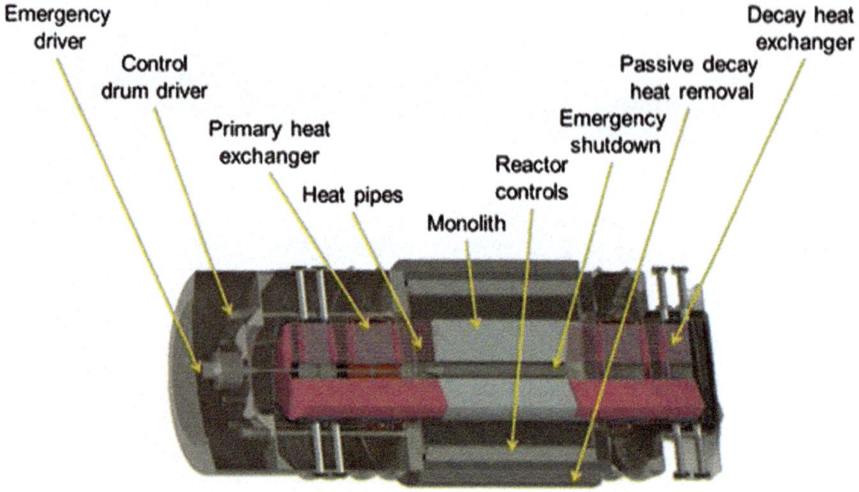

Fig. 2.16 Exposure of eVinci cross-section from different angle. (Source: Courtesy of Westinghouse Corporation)

Distinct from other small reactor designs, the eVinci is a heat pipe reactor, using a fluid in numerous sealed horizontal steel heat pipes to conduct heat from the hot fuel (where the fluid vaporizes) to the external condenser (where the fluid releases latent heat of vaporization) with heat exchanger. No pumps are needed to effect continuous isothermal vapor/liquid internal flow at low pressure. The principle is well established on a small scale, but here a liquid metal is used as the fluid and reactor sizes up to several megawatts are envisaged. Experimental work on heat pipe reactors for space has been with much smaller units (about 100 kWe), using sodium as the fluid. They have been developed since 1994 as a robust and low technical risk system for space exploration with an emphasis on high reliability and safety.

The eVinci reactors would be fully factory built and fueled. As well as power generation, process heat to 600 °C would be available. Units would have 5–10-year operational lifetime, with walkaway safety due to inherent feedback diminishing the nuclear reaction with excess heat, also effecting load following. Although eVinci is not considered as an Advanced Small Modular Reactor (AdvSMR), due to its small scale size is a good candidate for space based source of energy and military application of it due to its transportability and mobility feature as it is mentioned in above and falls into category of Micro Reactor for lack of better classification choice. As it is illustrated in Fig. 2.16, the eVinci micro reactor's innovative design is a combination of nuclear fission, space reactor technologies, and 50+ years of commercial nuclear systems design, engineering, and innovation. The eVinci micro reactor aims to create affordable and sustainable power with improved reliability and minimal maintenance, particularly for energy consumers in remote locations. The small size of the generator allows for easier transportation and rapid, on-site installation in contrast to large, centralized stations. The reactor core is designed to run for more than 10 years, eliminating the need for frequent refueling.

The key benefits of the eVinci micro reactor are attributed to its solid core and advanced heat pipes. The core encapsulates fuel to significantly reduce proliferation risk and enhances overall safety for the user. The heat pipes enable passive core heat extraction and inherent power regulation, allowing autonomous operation and inherent load following capabilities. These advanced technologies together make the eVinci micro reactor a pseudo "solid-state" reactor with minimal moving parts and its attributes were listed as before.

Heat pipes operate in a passive mode at relatively low pressures, less than an atmosphere. Each individual heat pipe contains only a small amount of working fluid, which is fully encapsulated in a sealed steel pipe. There is no primary cooling loop, hence no mechanical pumps, valves, or large-diameter primary loop piping typically found in all commercial reactors today. Heat pipes simply transport heat from the in-core evaporator section to the ex-core condenser in continuous isothermal vapor/liquid internal flow. Heat pipes offer a new and unique means to remove heat from a reactor core [2, 3].

Furthermore, Westinghouse is developing the eVinci micro reactor (Fig. 2.16), which ranges from 200 kW to 15 MW. Westinghouse says that it could develop and demonstrate the eVinci micro reactor in less than 6 years, owing to its size and "high technology readiness level of individual components." For now, it wants to develop a full-scale electrical demonstration unit by 2019 and then build an integral test with nuclear fuel, aiming for commercial deployment by 2024.

"These challenges require careful risk management and planning, but they are not considered showstoppers and their management is part of the Westinghouse eVinci reactor development program," the company says.

Bear in mind that the pioneer and the frontrunner in Small Modular Reactor (SMR) race, NuScale expects the US Nuclear Regulatory Commission (NRC) to approve its 60 MW design by September 2020, with commercial deployment slated in the mid-2020s. At the end of September, NuScale announced it selected Virginia-based BWX Technologies and spinoff of Babcock & Wilcox to start the engineering work to manufacture its SMR. However, the company is also evaluating several micro reactor concepts in the 1–10 MW range, and two of these light water micro reactor concepts—the single-unit NuScale Power Module (NPM) and the reduced-size NPM—could leverage current design and licensing efforts.

2.4 Heat Pipe Brief Summary

A heat pipe is a two-phase heat transfer device with a very high effective thermal conductivity. It is a vacuum tight device consisting of an envelope, a working fluid, and a wick structure. As shown in Fig. 2.17, the heat input vaporizes the liquid working fluid inside the wick in the evaporator section. The saturated vapor, carrying the latent heat of vaporization, flows toward the colder condenser section. In the condenser, the vapor condenses and gives up its latent heat. The condensed liquid returns to the evaporator through the wick structure by capillary

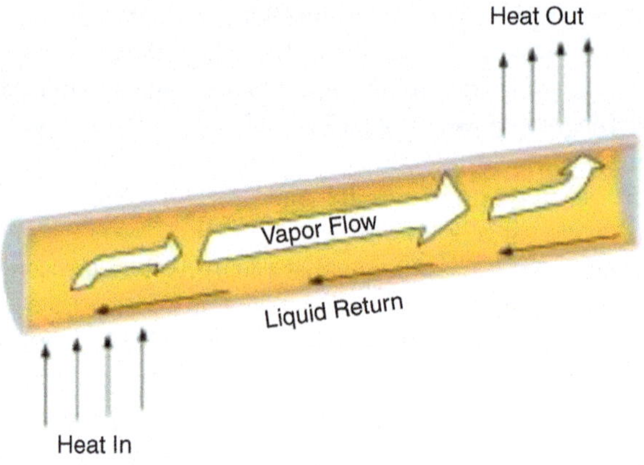

Fig. 2.17 A simple physical configuration of heat pipe

action. The phase change processes and two-phase flow circulation continue as long as the temperature gradient between the evaporator and condenser are maintained.

Heat pipes function by absorbing heat at the evaporator end of the cylinder, boiling and converting the fluid to vapor. The vapor travels to the condenser end, rejects the heat, and condenses to liquid. The condensed liquid flows back to the evaporator, aided by gravity. This phase change cycle continues as long as there is heat (i.e., warm outside air) at the evaporator end of the heat pipe. This process occurs passively and there is no external electrical energy required.

At the hot interface of a heat pipe a liquid in contact with a thermally conductive solid surface turns into a vapor by absorbing heat from that surface. The vapor then travels along the heat pipe to the cold interface and condenses back into a liquid—releasing the latent heat. The liquid then returns to the hot interface through either capillary action, centrifugal force, or gravity, and the cycle repeats. Due to the very high heat transfer coefficients for boiling and condensation, heat pipes are highly effective thermal conductors. The effective thermal conductivity varies with heat pipe length and can approach 100 kW/(m K) for long heat pipes, in comparison with approximately 0.4 kW/(m K) for copper.

Heat pipes employ evaporative cooling to transfer thermal energy from one point to another by the evaporation and condensation of a working fluid or coolant. Heat pipes rely on a temperature difference between the ends of the pipe and cannot lower temperatures at either end below the ambient temperature (hence they tend to equalize the temperature within the pipe) (Fig. 2.18).

Heat pipes have an envelope, a wick, and a working fluid. Heat pipes are designed for very long-term operation with no maintenance, so the heat pipe wall and wick must be compatible with the working fluid. Some material/working fluids pairs that appear to be compatible are not. For example, water in an aluminum envelope will develop large amounts of non-condensable gas over a few hours or days, preventing normal operation of the heat pipe.

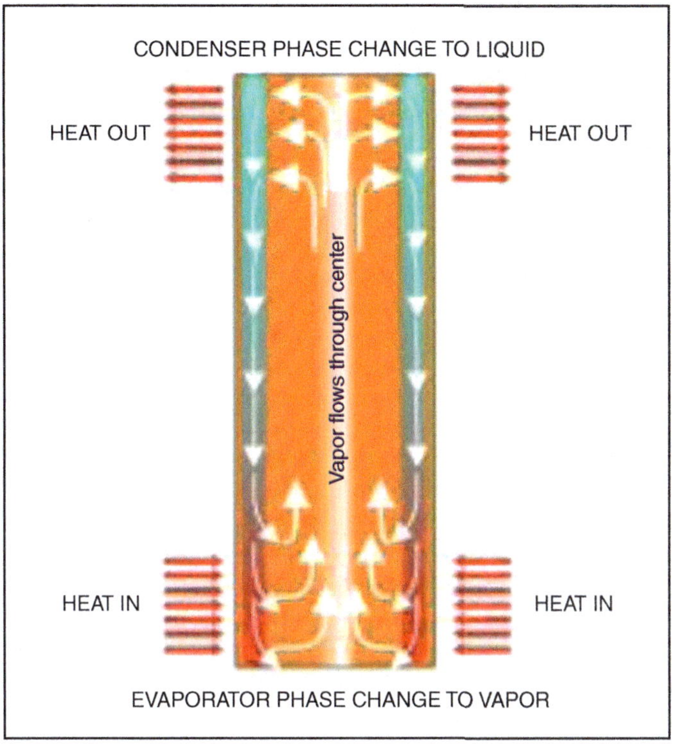

Fig. 2.18 Internal schematic heat pipe structure

2.4.1 *Heat Pipe Materials and Working Fluids*

Since heat pipes were rediscovered by George Grover in 1963, extensive life tests have been conducted to determine compatible envelope/fluid pairs, some going on for decades. In a heat pipe life test, heat pipes are operated for long periods of time, and monitored for problems such as non-condensable gas generation, material transport, and corrosion.

2.4.2 *Different Types of Heat Pipes*

In addition to standard, Constant Conductance Heat Pipes (CCHPs), there are a number of other types of heat pipes, including:

- Vapor Chambers (planar heat pipes), which are used for heat flux transformation, and isothermalization of surfaces.

- Variable Conductance Heat Pipes (VCHPs), which use a Non-Condensable Gas (NCG) to change the heat pipe effective thermal conductivity as power or the heat sink conditions change.
- Pressure Controlled Heat Pipes (PCHPs), which are a VCHP where the volume of the reservoir, or the NCG mass can be changed, to give more precise temperature control.
- Diode Heat Pipes, which have a high thermal conductivity in the forward direction, and a low thermal conductivity in the reverse direction.
- Thermosyphons, which are heat pipes where the liquid is returned to the evaporator by gravitational/acceleration forces.
- Rotating heat pipes, where the liquid is returned to the evaporator by centrifugal forces.

2.4.3 *Nuclear Power Conversion*

Grover and his colleagues were working on cooling systems for nuclear power cells for space craft, where extreme thermal conditions are encountered. These alkali-metal heat pipes transferred heat from the heat source to a thermionic or thermo-electric converter to generate electricity.

Since the early 1990s, numerous nuclear reactor power systems have been proposed using heat pipes for transporting heat between the reactor core and the power conversion system. The first nuclear reactor to produce electricity using heat pipes was first operated on September 13, 2012, in a demonstration using flattop fission.

In Nuclear power plant application, heat pipes can be used as a passive heat transfer system for performing as overall thermal hydraulic and natural circulation sub-system in an Inherent Shutdown, Heat Removal System (ISHRS) in the core (i.e., installed on top of the core doom) of nuclear reactor such as Molten Salt or Liquid Metal Fast Breeder type reactors, as a secondary fully inherent shutdown system loop acting like heat exchanger from safety point of view so the reactor never reaches to its melting point in case of accidental events [2, 3].

2.4.4 *Benefits of These Devices*

- High Thermal Conductivity (10,000–100,000 W/m K)
- Isothermal
- Passive
- Low cost
- Shock/vibration tolerant
- Freeze/thaw tolerant

2.4.5 Limitations

- Heat pipes must be tuned to particular cooling conditions. The choice of pipe material, size, and coolant all have an effect on the optimal temperatures at which heat pipes work.
- When used outside of its design heat range, the heat pipe's thermal conductivity is effectively reduced to the heat conduction properties of its solid metal casing alone—in the case of a copper casing, around 1/80 of the original flux. This is because below the intended temperature range the working fluid will not undergo phase change; and above it, all of the working fluid in the heat pipe vaporizes and the condensation process ceases.
- Most manufacturers cannot make a traditional heat pipe smaller than 3 mm in diameter due to material limitations.

2.4.6 Conclusion

Overall, a heat pipe is a heat transfer device that combines the principles of both thermal conductivity and phase transition to effectively transfer heat between two solid interfaces. See illustration in Fig. 2.19.

Phase-change processes and the two-phase flow circulation in the HP will continue as long as there is a large enough temperature difference between the evaporator and condenser sections. The fluid stops moving if the overall temperature is uniform but starts back up again as soon as a temperature difference exists. No power source (other than heat) is needed.

In some cases, when the heated section is below the cooled section, gravity is used to return the liquid to the evaporator. However, a wick is required when the evaporator is above the condenser on earth. A wick is also used for liquid return if there is no gravity, such as in NASA's micro-gravity applications [2, 3].

2.5 Miniaturization of New Generation of Nuclear Power Plants

After fall down of public opinions toward nuclear energy as source of generating electricity, companies like Westinghouse (W), Babcock & Wilcox (B&W) faced bankruptcy and reorganization and countries like Germany totally abandoned the idea of generating electricity from nuclear power in particular after nuclear disaster in Fukushima Daiichi nuclear disaster (2011) in Japan and divesting effect of it globally.

However, with an increase of population at the rate of almost 17% globally the demand for electricity is gone up and with reduction in Carbon Monoxide (CO) and Carbon Dioxide (CO_2) and greenhouse effect, we need nuclear energy in terms of

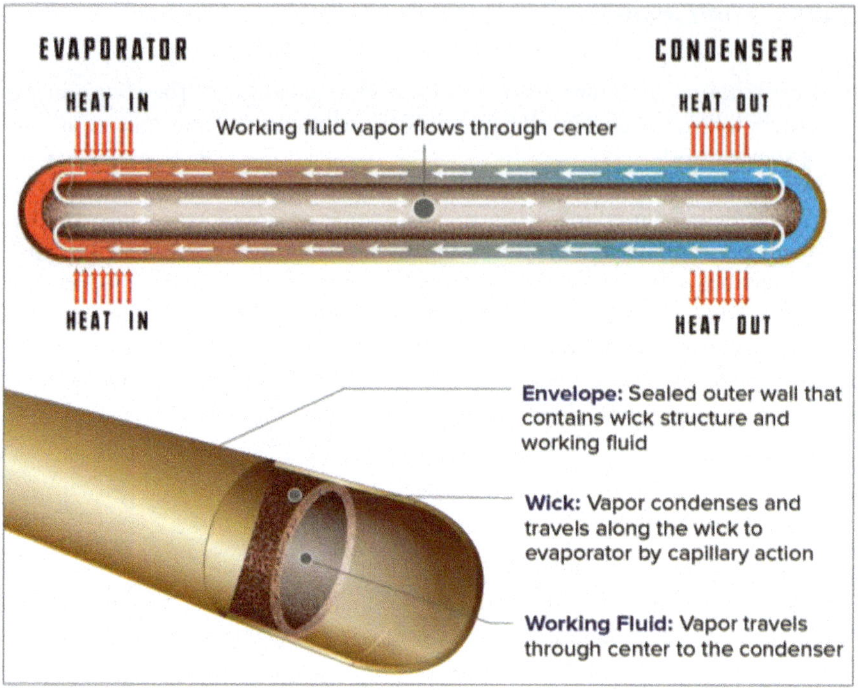

Fig. 2.19 Tope view depiction of heat pipe

Fission Process in present and near time as well as Fusion Process in long term when this technology gets more mature and will take off from research to production line. With proposed new Generation IV (GEN-IV) and more advanced toward safety and operation of these generations as well cost production effectiveness of these types of advanced reactors in the form of modularization comes the miniaturized version of these reactors into picture as well. Mobility and transportability of these reactors are of interest to our demand for electricity, so these rectors become more reasonable to own them and operate them, even in most remote area in the world. See Fig. 2.20.

With technology of combined cycle from thermodynamics point of view, the efficiency of these advanced reactors technologically are improved by producing a better thermal efficiency output [5–8].

Looking at existing technology of Generation III (GEN-III) of these Fission Nuclear Power Plant (NPP), these nuclear power plants are typically hulking structures made using billions of dollars of concrete and steel. But one company thinks that by going smaller, they could actually make nuclear power more affordable.

As illustrated in Fig. 2.21, NuScale Power based in Portland, Oregano, one of the industry pioneers in Small Modular Reactor (SMR), has submitted a design for what it describes as a "modular" nuclear power plant based on Light Water Reactor (LWR) technology. Each module is a self-contained 50-MW nuclear reactor loaded

Fig. 2.20 New generation of power plant are transportable

with standard uranium reactor fuel. Modules would be assembled at a factory-like facility and then delivered to power utilities and other clients and their locations. NuScale says the uranium fuel in each of its modules would be housed in a special containment vessel that would be submerged in a pool of water, an added safety feature.

Other companies such as Holtech, General Electric/Hitachi combined and Westinghouse are suggesting very similar approach to such modularization and miniaturization as well. As we stated, Westinghouse with their eVinci Nuclear Micro Reactor is leading such effort.

"Miniature" in nuclear terms is still pretty big. The modules are small enough to fit on flat-bed trucks, but they would stand nearly nine stories tall. Moreover, a power plant would probably require several modules hooked together like giant batteries. Of course, they would need to be operated by professional nuclear engineers.

Historically, it is the first time commercial nuclear power plant manufactures are suggesting and submitting an application to the Nuclear Regulatory Commission (NRC) to build a small modular type reactors.

The traditional concept of the nuclear power plant conjures images of massive cooling towers, spent fuel pools, cavernous underground tunnels, and worst-case scenarios. But it does not have to be that way. US nuclear power companies submitted formal plans to build pre-assembled mini-reactors in the USA that are small enough to be transported by boat or on the back of trucks as illustrated in Figs. 2.20 and 2.22.

https://www.seeker.com/a-nuclear-energy-company-wants-to-build-americas-first-small-modular-r-2212178795.html

These generation of nuclear power plants design are a radical departure from the tradition generation (i.e., GEN-III) would be safer and more cost-effective to build and offers more advantages from Total Cost of Ownership (TCO).

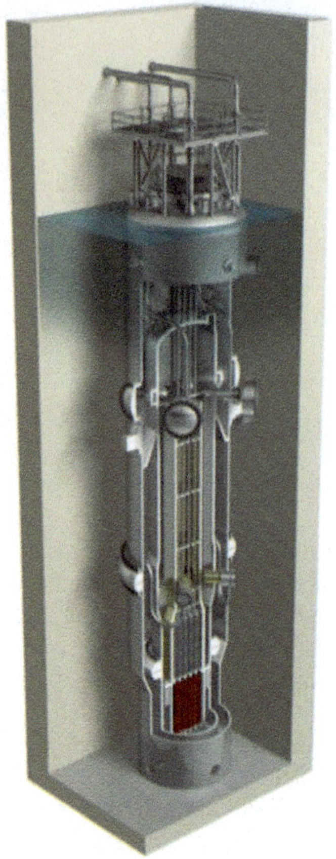

Fig. 2.21 The NuScale uranium fuel module concept

Few advantages that we can list here are:

- Proponents of these advanced small modular reactors say they will be easier to build and more flexible in terms of where they can be located than the larger kind. The word "modular" refers to how they will be built in factory-like settings, ready for hauling either fully assembled or in easily connected parts by truck, rail, or sea.
- These reactors can potentially power rural towns, industrial plants, mountainous areas, and military bases, as well as urban districts and ports. Small modular reactors may also prove handy for industrial uses.
- Small modular reactors will differ from the smaller reactors already deployed because of their new technologies. These advances are intended to make it less likely or even impossible for them to melt down or explode, as happened during Japan's Fukushima disaster.
- The power plants where these small reactors will be located will have added protections against sabotage and the theft of radioactive material. For example, they may be equipped with cooling systems that continue working even if no

Fig. 2.22 Using barge/boat to transport small modular reactor

operators are present and all electric power is lost. In many cases, the entire reactor and steam-generating equipment will be below ground to safeguard these facilities during natural disasters like the earthquake and tsunamis that led three Fukushima Daiichi reactors to melt down.

- Like renewable energy, nuclear power emits no carbon. And compared to wind and solar power, which are intermittent sources, or hydropower, which is affected by seasonal changes and droughts, it operates all the time and has a much smaller footprint.
- As a result, small modular reactors could be paired with renewable sources as a substitute for coal-fired or natural gas plants. Yet they will probably have to compete with advanced energy storage systems for that market.

As we stated in above, each module uses less uranium fuel, making a large-scale meltdown far less likely. The fuel would be housed in a special containment vessel that would be submerged in a pool of water, an added safety feature. And rather than using pumps of the sort that failed during the Fukushima nuclear disaster, the reactor would circulate the water using natural convection. The company maintains the design is simpler and safer than existing reactors.

NuScale formally completed its design submission to the US Nuclear Regulatory Commission (NRC) on Thursday. The 12,000-page application will now undergo a lengthy review by the NRC, which must approve the design before construction can begin.

Moving forward with the idea of SMRs and Micro Reactor is the cost-effectiveness challenges of such miniaturized reactors. The cost-effectiveness of utilizing them on a large scale. Although SMRs are cheaper initially due to a lower cost per unit, the most obvious drawback is the increased running costs [9]. Since these reactors are smaller and produce less energy per unit, due to economies of scale, power output decreases while other costs stay constant [10]. This is probably the most substantial reason for why SMRs have not yet emerged in a widespread context.

Other challenges include dealing with the waste products of SMRs; however, this is the same for full-scale nuclear reactors as well. Additionally, people have mentioned the existence of SMRs could be dangerous for terrorist activity; however, I do not foresee this being as significant of a problem as some people have predicted, considering even SMR sites have security and are managed closely [11].

Although long-term costs remain a barrier to SMRs becoming a new standard in the energy sector, this may change within the next few years. In terms of cost structures, the Andlinger Center review finds that "If the number of small plants constructed becomes large enough, if 'learning' is strong enough, and if the diseconomies of small scale are weak enough, five 200-megawatt reactors could become cheaper than one 1,000-megawatt plant" [10].

On-site assembly will be faster and cheaper as well, since it is much less complex than building reactors from the ground up (and down). Once on site, the 76-ft tall reactor is lowered into an underground containment vessel, itself submerged in water within a vertical, steel-and-concrete reactor pool. The entire underground structure is then capped with a shield cover, with the power plant facility constructed on top. This kind of modular power plant development has many advantages, according to proponent of these reactors as we stated them in above.

Whether these advantages materialize, obviously, remains to be seen once these reactors are deployed. Some experts are skeptical of the industry's promises and expectations.

Although small modular reactors are designed to produce less radioactive waste than standard, bigger reactors for the same amount of power, the issue of where to safely dispose of nuclear waste remains unresolved.

Small modular reactors face other challenges, some of their own making. Strong interest in the potential global market has led many companies to propose their own individual reactor designs. In my opinion, there are already too many versions out there. Before long, a shakeout will occur.

It is also unclear what small modular reactor-generated power will cost. That will probably remain the case for at least the next 10–15 years, until a few designs are actually built and operating.

Some experts foresee small modular reactors penciling out at levels that could be higher than for full-sized reactors which generally cost more to build and operate than other options, like natural gas, for the same amount of power. NuScale, however, predicts that its SMRs will be more competitive than that in terms of their cost.

These are encouraging speculation to suggest that SMRs do have a bright future for progress in the energy sector. Moreover, there are two highly compelling potential applications for SMRs. The first situation that may be ideal for SMRs is "in groups, where several small reactors are an alternative to one large one." Secondly, SMRs may be applicable when used "individually, in remote, isolated locations where a large reactor is unsuitable and not considered feasible either" [10].

2.6 Nuclear Micro Reactor and Military Application

As we have learned in the first two chapters so far, SMRs are already being utilized by multiple parties, including "on nuclear submarines and in some developing countries such as India and Pakistan" [12]. Some major upsides of SMRs compared to larger nuclear reactors is that they are more transportable, require less uranium fuel (which could potentially lead to fewer meltdowns), and are more affordable at initial market prices. One of the major advantages to SMR technology is the initial economic benefit. While large nuclear reactor sites are extremely costly and difficult to finance, SMRs are more feasible and thus open up the opportunity of harnessing nuclear energy for more parties globally. To illustrate this matter, consider France for a moment.

A French energy company, EDF, had plans to build new large reactor sites in France and Finland. However, due to potential safety concerns, "the plans went billions of euros over budget" [12]. The problem lies in the fact that large nuclear reactors take more time to build and check safety features. Thus, SMRs could help overcome this barrier of financial and time pressures. Evidently, SMRs have already begun to infiltrate our world in tangible ways. However, SMRs are not as widespread as one might predict considering their many advantages. As governments and private entities begin to adopt SMRs and harness the capability of SMRs to produce "cleaner energy," it is also important to consider the potential downsides and dangers associated with SMRs.

With more than 70 advanced nuclear reactor projects in various stages of development in the USA alone, there is exciting growth in this field. "Micro reactors" are one class of these innovative technologies, whose particular attributes hold out special promise to the nation's largest energy user—the US military.

When it comes to the military application of these small modular reactors, the US military is in quest of ting road mobile nuclear reactors that can fit in a cargo airplane such as C-17 or C-5 Galaxy or get on a truck such as the one illustrated in Fig. 2.23.

The power demands to sustain American military operations are only increasing, but small nuclear power plants could present new problems. The US military's secretive Strategic Capabilities Office (SCO) is asking for potential vendors to submit proposals for small mobile nuclear reactors to help meet ever-growing demands for power during operations in remote and austere locations. This request for information comes as the US Army, in particular, is looking to extend the amount of time its units can operate independent of established supply chains, but portable nuclear power could introduce new risks to the battlefield.

Fig. 2.23 Transporting Nuclear Micro Reactor via Military Truck

Concerns are growing throughout the US military about the potential difficulties in rapidly deploying large amounts of personnel and equipment into a theater of operations under fire during a major conflict and whether there will be any bases of operation to support them once they get there. Now, the US Army says it is looking for ways to ensure that individual brigade combat teams will have supplies, especially fuel and water, to be able to keep fighting for up to a week without a guaranteed supply chain.

As the US military's primary focus shifts to potential high-end conflicts with adversaries such as Russia and China, senior US Air Force (USAF) leaders are warning that the service must be prepared to deploy to a region and quickly set up new bases. Otherwise, the service risks seeing its operations hampered, if not brought to a halt entirely for at least some amount of time, if an enemy has destroyed or otherwise renders established facilities unusable in the opening stages of a major war.

SCO first announced that they were looking for "information on innovative technologies and approaches" relating to a possible future "small mobile nuclear reactor prototype design" on FedBizOpps, the US government's main contracting website, on January 18, 2019. The organization posted an amended version of the notice, which outlines a "multi-phase prototype project" as part of what it is calling Project Dilithium, 4 days later.

"Energy usage during contingency operations will likely increase significantly over the next few decades," the latest version of the request for information explains. "The modern operational space has amplified the need for alternative energy sources to enable mobility in forward land based and maritime military operations." As its concept is artistically drawn as depicted in Fig. 2.24.

Note that FOB, "Free-on-Board," is a term in international commercial law specifying at what point respective obligations, costs, and risk involved in the delivery of goods.

Moreover, a diagram from a previous US military report depicting a potential concept of operations for deploying a mobile nuclear reactor.

SCO basic requirements envision a reactor that can generate between 1 and 10 MW of energy, less than the average output for even a small research reactor and weigh less than 40 tons. The final design would need to be portable by semi-trailer truck, ship, or a US Air Force C-17A Globemaster III cargo plane. See Fig. 2.25.

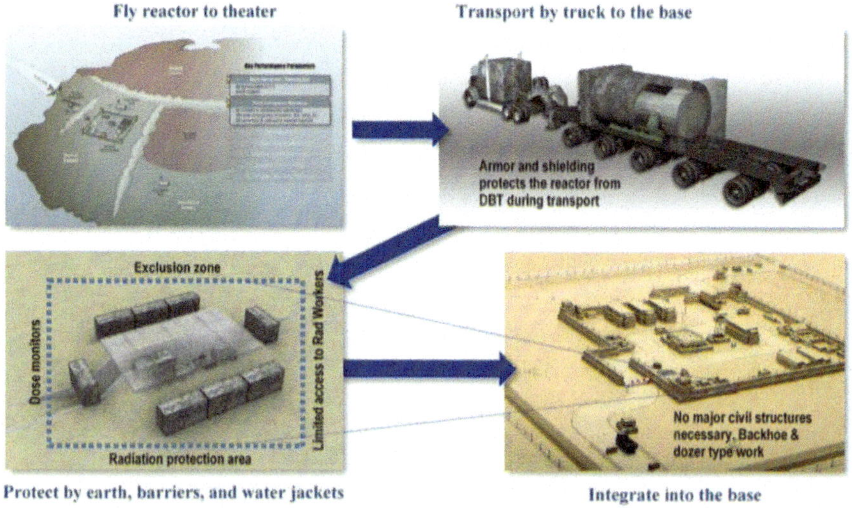

Fig. 2.24 Concept of operations, transport to theater and Free-on-Board (FOB)

The reactors would produce 1–10 MWe of electrical power, weight less than 40 tons, be capable of being set up and running after delivery by a C-17 in less than 72 h; and, run more or less unattended for up to 3 years.

If the Army is successful in developing and deploying these mini reactors, the supply chain created by producing them could be capitalized by current and future developments of small modular reactors.

Additionally, assuming the US Army produces enough of them, that cost and schedule data from the line might finally answer the question of how many orders a commercial SMR developer needs to start up its own production line. Note the Army is unlikely to want to share a production capability needed to assure tactical readiness at forward bases.

The testing and safety review processes for the military reactors would also provide test cases for use by commercial reactors and also inform the NRC about getting smart on these types of reactors. According to the Army Times report test efforts are expected to take place at the Idaho National Laboratory (INL) and an interest has been announced by INL as well.

Note that: The specifications for the acquisition of these mini-reactors, sometimes referred to incorrectly as "***nuclear batteries,***" was published in the Federal Business Opportunities aka FedBizOps on 22 January 2019. The announcement came with a detailed list of operational and physical requirements. This announcement is a "Request for Information (RFI)." A full blown Request for Proposals (RFP) could be issued later this year.

Note to Readers—4 March 2019—*the correct designation of the issuing office of the RFI is the Office of the Secretary of Defense, Strategic Capabilities Office aka OSD/SCO as an acronym. References to the US Army actually refer to this part of the Pentagon.*

Fig. 2.25 Cargo Plane C-17 photo

Furthermore, the RFI says that the Defense Department could pick three prototypes for its phase I portion of development. That would require prototype designs and other plans. That phase would go for 9–12 months, according to the post. The Idaho National Laboratory is cited as offering an estimate that testing and demonstrations could begin as early as 2021.

There are a number of potential concepts already in various stages of development that could meet SCO's requirements. The US Department of Energy's own Los Alamos National Laboratory (LANL), in cooperation with the Westinghouse power company, has been working on one design called MegaPower (i.e., Fig. 2.26) for some time now. Westinghouse is separately working on its own eVinci micro reactor design.

In September 2018, officials at Los Alamos National Laboratory released details of work taking place there to build a very small reactor to meet the military's needs.

In the LANL micro reactor, the nuclear fuel is HALEU. The fuel is encapsulated in a solid-steel monolith to form a sub-critical nuclear core, which is surrounded by a neutron reflector and contains a simple shutdown rod that allows the core to go critical on demand.

The thermal energy created by the fission reactions is efficiently removed from the metal core by high temperature, alkali-metal heat pipes, a technology in wide use both on earth and in space since the 1970s. Thermal energy is converted into electrical energy—and power is delivered.

As part of MegaPower reactor design, this new micro reactor sprang from Los Alamos' work to develop a small nuclear reactor for NASA that might someday power a colony on Mars or the Moon. Called Kilopower, the Mars reactor is about

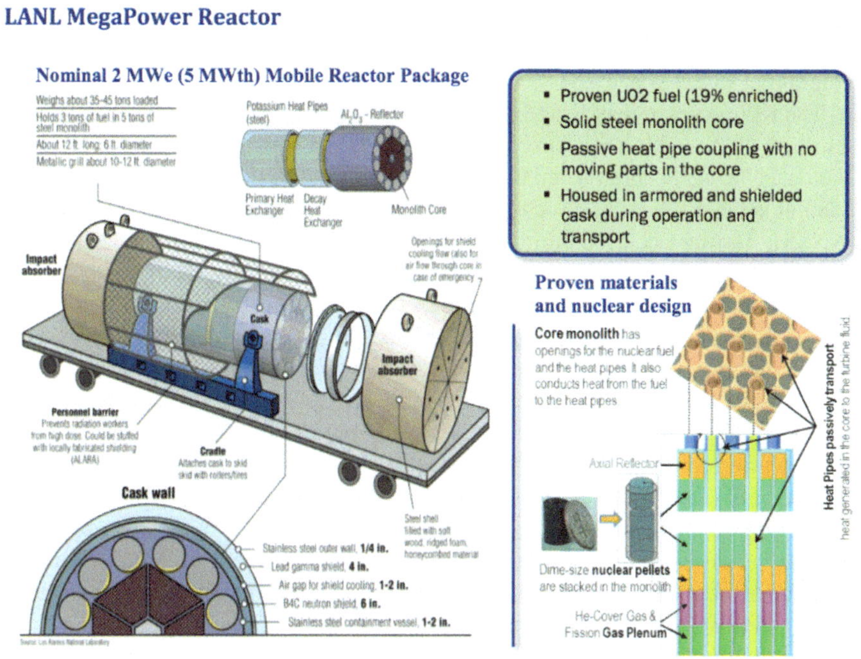

Fig. 2.26 Los Alamos National Laboratory MegaPower Reactor Systems Design. (Source: LANAL Courtesy)

the size of a roll-aboard suitcase and provides 1–10 kW of electric power. A key feature is self-regulation: the Kilopower reactor relies on inherent physics phenomena to naturally regulate its power output to meet power needs. Passive heat pipes drive multiple engines to generate electricity. Testing at the National Criticality Experiments Research Center at the Nevada National Security Site demonstrated a smaller version of the entire system.

The Kilopower project is a near-term technology effort to develop preliminary concepts and technologies that could be used for an affordable fission nuclear power system to enable long-duration stays on planetary surfaces (see Fig. 2.27).

After successful completion of the Kilopower Reactor Using Stirling Technology (KRUSTY) experiment in March 2018, the Kilopower project team is developing mission concepts and performing additional risk reduction activities to prepare for a possible future flight demonstration. Such a demonstration could pave the way for future Kilopower systems that power human outposts on the Moon and Mars, enabling mission operations in harsh environments and missions that rely on In situ Resource Utilization to produce local propellants and other materials.

The Kilopower project is part of NASA's Space Technology Mission Directorate's Game Changing Development (GCD) program, which is managed by NASA's Langley Research Center. The Flight Opportunities program funded the parabolic and suborbital flights that matured the Kilopower technology's titanium water heat pipes by exposing it to space-relevant environments through the use of commercial reusable suborbital launch vehicles. The project will remain a part of the GCD program with the goal of transitioning to the Technology Demonstration Mission program in Fiscal Year 2020.

The Kilopower project team is led by NASA's Glenn Research Center in partnership with NASA's Marshall Space Flight Center; the Department of Energy's (DOE) National Nuclear Security Administration (NASA); and NNSA's Los Alamos National Laboratory, Y-12 National Security Complex, and Nevada National Security Site (NNSS). Testing occurred from November 2017 to March 2018 at the National Criticality Experiments Research Center in the Device Assembly Facility at the NNSS.

Keep in mind that, "**There is not much the**" US military does that is more dangerous than trucking fuel through a war zone. In 2009, the Army found that one soldier died for every 24 fuel convoys in Afghanistan. So, if a better way could be found to generate electricity at remote bases—that is what most of the fuel is used for—it could greatly reduce the risks to our military (Fig. 2.28).

A solution could be a new micro-nuclear reactor being developed by Los Alamos National Laboratory and the Westinghouse power company. Built around heat pipe technology [2, 3], this inherently safe micro reactor has no cooling water or pumps that can fail, uses passive regulation systems so that it cannot melt down, and can generate at least 1 MW of safe, reliable power for 10 years or more. A megawatt is enough electricity for roughly a military brigade, some 1500–4000 soldiers.

Pay attention to the fact that the Pentagon has been interested in small nuclear reactors for nearly a decade, as robust and reliable power sources for military installations within the territorial United States and overseas. The very small "micro reactor" designs now emerging from companies like General Atomics, NuScale Power LLC, Oklo Inc., Westinghouse Electric Co., and X-energy LLC are particularly well aligned with DOD's needs for energy security and resilience.

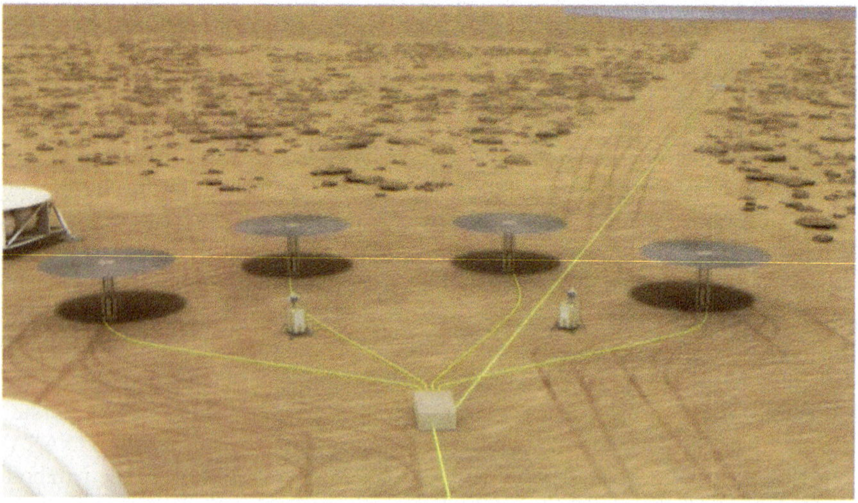

Fig. 2.27 Conceptual reactor image on planetary surface

Micro reactors are a source of resilient energy that can enable a wide range of installations to enhance their range, endurance, agility, and mission assurance. Micro reactors are being designed with island-mode operations, black-start capabilities, an ability to protect against severe natural phenomena as well as man-made physical and cybersecurity threats, and to operate for several years without the need to shut down for refueling.

Such strong interest for Small Modular Reactors and Nuclear Micro Reactors can even power remote communities, both *Military* and *Civilian* (i.e., Fig. 2.29) and working closely with the reactor industry and vendors and with relevant offices at the US Department of Defense (DOD) and Department of Energy (DOE), the Nuclear Energy Institute (NEI) in release of its published report showed a Road Map laying out the actions needed and to take and to ensure the successful deployment of these new generations of a First-of-a-Kind micro reactors at a domestic defense installation such as INL by the end of 2017.

NEI's technical report, "Road Map for the Deployment of Micro Reactors for U.S. Department of Defense Domestic Installations," [13] provides guidance and input for the pilot program. The road map outlines the timelines and challenges for deploying the first micro reactor at a domestic DOD facility and recommends actions to ensure its installment by December 31, 2027, as prescribed in the NDAA.

Among NEI's recommended actions for DOD, DOE, and industry are:

- DOD should identify the host installation and site requirements, perform an assessment of the designs, and enter into a contract or agreement with a commercial entity by the end of 2019.
- DOE should provide high-assay low-enriched uranium (HALEU) by the end of 2022. This is uranium enriched in fissile uranium-235 to a slightly higher level than that used by current commercial nuclear reactors and is needed for some micro reactor designs.
- Developers should sustain the development of micro reactor designs to enable deployment within the program timeline, including designing for manufacturability, constructability, and operability, and optionally entering into private–public partnerships with DOE.

Fig. 2.28 Trucking nuclear reactor energy through a war zone

Fig. 2.29 Conceptual remote civilian community

- DOD should work closely with the industry and the US Nuclear Regulatory Commission to identify and resolve unique regulatory issues associated with micro reactors.
- Industry should immediately begin working with the NRC to explore options for accelerating the review schedule for micro reactors.

While the road map focuses on domestic defense installations, micro reactors are equally suitable for forward operating bases overseas. For these applications, the Pentagon also is interested in micro reactors for their potential to eliminate the need for fuel resupply.

The SCO's requirements demand a design that is incapable of melting down, does not require constant monitoring for otherwise safe operation, and does not present an immediate health hazard. LANL and Filippone and Associates LLC both claim that their designs meet these requirements through inherent safety features in their respective designs.

But even if the reactor itself cannot catastrophically fail, something that may be a tall order to ensure in austere conditions regardless of the design, powering remote and austere bases with nuclear power could run other risks. If hostile forces end up destroying the reactor, it could potentially lead to the hazardous dispersal of radioactive material. See Fig. 2.30.

The Department of Defense wants a portable nuclear reactor the size of a main battle tank that is capable of being lifted to overseas hot spots. The reactor would provide megawatts of power for US forces, providing juice for everything from Xboxes to directed energy weapons. A mobile reactor would also make the military less reliant on diesel fuel for electrical generators, which in some cases must be sent along dangerous resupply lines.

According to the Federal Business Opportunities website, the Department of Defense's Strategic Capabilities Office has officially put out a request for information for a "Small Mobile Nuclear Reactor." Based at the Pentagon, the Strategic Capabilities Office (SCO) spearheads the development of cutting edge military technology.

Unlike the Defense Advanced Research Projects Agency, which is more concerned with developing technologies, the SCO is more focused on deliverable weapons systems and other technologies, even such as the Small Mobile Nuclear Reactor (SMNR) that is described above.

Bottom line is that, the US Army or military wants its Brigades to be able to fight anytime anywhere around the globe to defend interest of the USA for at least entire week without resupply and have source of generating electricity free of any chemical fuel, such as diesel or gasoline and that is these new generations of first kind micro reactors are looking very appealing.

The service is worried that units have grown dangerously reliant on logistics chains that might not exist during a major conflict.

2.6.1 Department of Defense Requirements

Resilience is a primary requirement for DoD. Resilience is defined in the 2018 NDAA as "the ability to avoid, prepare for, minimize, adapt to, and recover from anticipated and unanticipated energy disruptions in order to ensure energy availability and reliability sufficient to provide for mission assurance and readiness, including task critical assets and other mission essential operations related to readiness,

Fig. 2.30 Military personnel conducting radiological reconnaissance mission in remote areas

and to execute or rapidly reestablish mission essential requirements." The 2018 NDAA also defines energy security as "having assured access to reliable supplies of energy and the ability to protect and deliver sufficient energy to meet mission essential requirements." Energy resilience must be achieved in a manner that does not negatively impact the mission. Generation sources that minimize usage of land, air, and water resources, including site access for fuel receipt and storage, will enhance the installation's ability to focus on mission-critical activities.

Currently, DoD installations rely almost entirely on the grid, which is highly vulnerable to prolonged outage from a variety of threats, placing critical missions at unacceptably high risk of extended disruption. Backup power is often based on diesel generator sets with limited on-site fuel storage, undersized for new Homeland defense missions, not prioritized to critical loads and inadequate in duration and reliability. This is according to a Defense Science Board Task Force on DoD Energy Strategy that also found "critical national security and Homeland defense missions are at an unacceptably high risk of extended outage from failure of the grid" [14].

DoD Instruction 4170.11, Installation Energy Management, provides policy direction to ensure energy resilience at military installations. Each military branch appears to address energy resilience differently, for example, the Army contextualizes resilience in terms of energy security, while the Air Force contextualizes resilience in terms of energy assurance. Efforts to enhance resilience also appear to focus along the lines in which resilience has been contextualized [15].

While the detailed scenario and capability requirements necessary to provide resilience at DoD installations are not known and may be unique to each facility, there are a number of considerations that appear to be universally important. These include minimization of outages due to maintenance and failures, being robust against natural and man-made threats, the ability to quickly restore critical operations, and having secure access to the source, delivery and on-site storage of fuel. More detailed requirements for resilience for Air Force Installations are contained in a 2017 RAND report that is published [16].

Micro reactors offer the ability to assure energy resilience at DoD installations. Micro reactors can operate 24 hours a day, 7 days a week, 365 days a year, and are being designed to operate for several years without the need to shut down for refueling. Micro reactors are being designed to protect against severe natural phenomena as well as man-made physical and cyber security threats, and many are being designed with the ability to operate in island-mode and include black-start capabilities. Micro reactors are also being designed to be able to adjust output to meet changes in demand.

Micro reactors will be able to provide heat and other products, such as desalinated water and hydrogen that can meet the needs of DoD installations. Micro reactors can also enhance DoD's use of new technologies, such as advanced computing, "big data" analytics, artificial intelligence, autonomy, robotics, directed energy, hypersonic, and biotechnology. The very technologies that ensure the USA will be able to fight and win the wars of the future [17].

Other requirements such as cost, environmental impacts, power characteristics, and site parameters, including land area, water sources, seismicity, precipitation, and wind speed, are not known and are likely to be site specific.

2.7 Nuclear Micro Reactor Influencing Future Space Explorations

Historically, since our time of moon walking with National Nuclear Security Administration (NASA) behind it and trying to explore space by going beyond moon, we had a lot of interest to integrate source of nuclear energy as part of population system so we can travel faster and further in space.

You may not associate space travel with the Energy Department. But you should—because nuclear power systems developed here have made dozens of truly amazing interplanetary research missions possible with Department of Energy (DOE) being involved since its their charter of the work around nuclear power energy.

The Energy Department's Office of Space and Defense Power Systems and its predecessors, in tandem with the National Labs and private industry partners, have developed and provided radioisotope power systems to NASA for use in numerous long-term missions, from Voyagers 1 and 2 to the Mars rovers (i.e., Fig. 2.31).

These compact, reliable systems provide basic mission fuel and keep critical spacecraft components warm enough to function in the cold, dark reaches of deep space and allows us to go about crossing into uncharted territory is that you may not know when, exactly, you have crossed into it. No one needs to tell that to the *Voyager 1* spacecraft, which is currently at the center of a controversy about where the solar system ends and interstellar space begins, see Fig. 2.32.

Although relatively simple, these systems have powered some of the most successful and inspiring missions in US space program history. Explore our interactive timeline of these missions, above, and read on to learn more about the "space batteries" that made them a reality.

On around March of 2011, the Mars Science Laboratory rover, Curiosity, launched from Cape Canaveral with the most advanced payload of scientific gear ever used on the red planet.

Its mission: to investigate whether the Gale Crater on Mars has ever offered environmental conditions that support the development of microbial life.

We should be proud to say that this mission is made possible by nuclear space power systems developed by the Energy Department. This year marks the 50th anniversary of nuclear-powered space exploration. Exploration such as the mission to Mars certainly shows a need for nuclear energy as population system on-board so it provided us more reliable and longer life cycle for fuel to go further and deeper in space.

On Curiosity, the Department's Multi-Mission Radioisotope Thermoelectric Generator (MMRTG) will provide continuous power to the rover and effective operating temperatures for its 11 scientific instruments. This is particularly important on

Fig. 2.31 Image of Voyager 1 probe in space. (Source: NASA)

Fig. 2.32 Deep space exploration. (Source: NASA)

Mars where reliable power is critical to maintain operation of the rover in the planet's dusty environment and during the cold Martian nights and frigid winter.

The MMRTG, which was constructed, assembled, and tested by the Energy Department and the Idaho, Oak Ridge, Los Alamos, and Sandia National Laboratories, uses heat produced by the natural decay of plutonium-238 to generate 110 W of electricity. See Fig. 2.33.

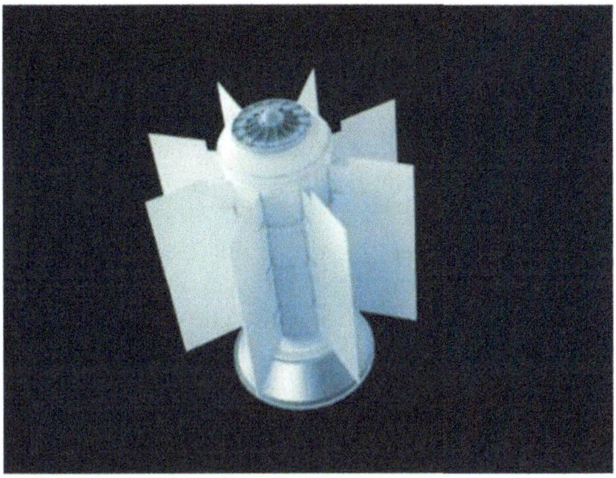

Fig. 2.33 MMRTG package. (Source: NASA)

The Curiosity missions investigate whether the Gale Crater on Mars has ever offered environmental conditions that support the development of microbial life.

These power systems have been used for space missions since 1961, including on the Apollo and Viking missions; the Galileo and Cassini spacecraft; and the Voyager probes.

When we are looking at augmentation and integration of nuclear reactor power as part of space explorer such as Voyager 1 and 2, then the size is the matter during lift and launch of such system into the space and mission that we are curious to explore. Nuclear Micro Reactor (NMR) is one way to go, and there exists a lot of interest around such technology by NASA scientist and nuclear engineers developing it in order to make space worthy.

2.7.1 Power From Plutonium

Despite what you see in movies and TV shows, there are only two practical ways to supply electrical power for multi-year space missions: the sun's rays or heat generated by natural radioactive decay. Radioisotope power systems—which directly convert heat generated by the decay of plutonium-238 into electric power—use the latter and are essential for long missions to distant parts of the solar system, where solar-powered space travel may be impractical or impossible.

Plutonium-238 (Pu-238) works well as a space power source for several reasons. It has a half-life of 88 years, meaning it takes that long for its heat output to be reduced by half. It is stable at high temperatures; can generate substantial heat in small amounts; and emits relatively low levels of radiation that is easily shielded, so mission-critical instruments and equipment are not affected. This type of plutonium is different than those used for nuclear weapons or nuclear power plant reactors.

In a Radioisotope Power System (RPS), commonly called a "space battery," the plutonium is processed into a ceramic form—similar to the material in your morning coffee mug. Just like a shattered mug, it breaks into large chunks instead of being vaporized and dispersed, preventing harm to people and the environment in the unlikely event of a launch or reentry accident. For more than 50 years, every radioisotope power system launched into space has worked safely and exactly as designed.

Technology involved in power from plutonium-238 as depicted in Fig. 2.34 is an overall image of this fuel element, which provides a General Purpose Heat Source (GHPS) as a module. The GPHS module provides steady heat for a radioisotope power system. See Fig. 2.35.

Radioisotope power systems, or RPS, provide electricity and heat that can enable spacecraft to undertake scientific missions to environments beyond the capabilities of solar power, chemical batteries, and fuel cells.

RPS are sometimes referred to as a type of "nuclear battery." While some spacecraft, like Cassini, do run their systems directly off of their RPS, others like the Mars Science Laboratory rover can use the RPS to charge batteries and run their systems and instruments off of stored battery power. In either case, the RPS is attached directly to a spacecraft, much like a power cord being plugged in.

These technologies are capable of producing electricity and heat for decades under the harsh conditions of deep space without refueling. All of these power systems, flown on more than two dozen NASA missions since the 1960s, have functioned for longer than they were originally designed.

The RPS used to power NASA spacecraft are supplied by the US Department of Energy (DOE). NASA and DOE continue to collaborate on maintaining and developing several types of RPS.

The General Purpose Heat Source module, or GPHS, is the essential building block for the radioisotope generators used by NASA. These modules contain and protect the plutonium-238 (or Pu-238) fuel that gives off heat for producing electricity.

Fig. 2.34 A Pu-238 fuel pellet protected via indium classing. (Source: NASA)

The fuel is fabricated into ceramic pellets of plutonium-238 dioxide ($^{238}PuO_2$) and encapsulated in a protective casing of iridium, forming a fueled clad. Fueled clads are encased within nested layers of carbon-based material and placed within an aeroshell housing to comprise the complete GPHS module.

Each GPHS is a block about 4 × 4 × 2 in. in size, weighing approximately 3.5 pounds (1.5 kg). They are nominally designed to produce thermal power at 250 W at the beginning of a mission and can be used individually or stacked together.

Modules have been subjected to extreme testing conditions that significantly exceeded the intensity of a wide range of potential accidents. Such tests have included simulating multiple reentries for a single module through Earth's atmosphere, exposure to high temperature rocket propellant fires, and impacts onto solid ground.

The enhanced GPHS modules used in the latest generation of radioisotope power systems incorporate additional rugged, safety-tested features that build upon those used in earlier generations. For example, additional material (20% greater in thickness) has been added to the graphite aeroshell and to the two largest faces of the block-like module. These modifications provide even more protection to help to contain the fuel in a wide range of accident conditions, further reducing the potential for release of plutonium-238 that might result.

2.7.2 *Radioisotope Power Systems Type in Radioisotope Thermoelectric Generator*

A Radioisotope Thermoelectric Generator (RTG) provides power for spacecraft by converting heat generated by the natural radioactive decay of its fuel source, plutonium dioxide, into electricity using devices called thermocouples. RTGs have no moving parts. The latest RPS to be qualified for flight, the Multi-Mission Radioisotope Thermoelectric Generator (MMRTG) is powering the Mars Science

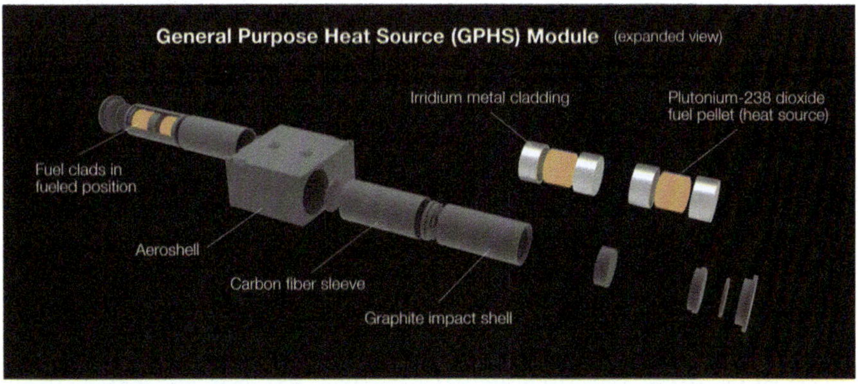

Fig. 2.35 General Purpose Heat Source (GPHS) module expanded view. (Source: NASA)

Laboratory rover, Curiosity, which landed on Mars in August 2012. The MMRTG continues to perform as designed, providing both power and heat for the rover.

As we stated before, the MMRTG is designed to be used in the vacuum of space as well as within the atmosphere of Mars. See Fig. 2.36.

Thermocouples are common in everyday items that must monitor or regulate their temperature, such as air conditioners, refrigerators, and medical thermometers. The principle of a thermocouple involves two plates, each made of a different metal that conducts electricity. Joining these two plates to form a closed electrical circuit while keeping the two junctions at different temperatures produces an electric current. Each of these pairs of junctions forms an individual thermocouple. In an RTG, the radioisotope fuel heats one of these junctions while the other junction remains unheated and is cooled by the space environment or a planetary atmosphere.

2.7.3 Radioisotope Heater Unit (RHU)

A Radioisotope Heater Unit (RHU) employs a small, pencil eraser-sized pellet of plutonium dioxide to generate heat for spacecraft structures, systems, and instruments, enabling their successful operation throughout a mission. Some missions employ just a few RHUs for extra heat, while others have dozens. NASA has also studied the potential for using the same small fuel pellet in an RHU to power a compact system that could provide a few dozen milliwatts of electrical power.

2.8 Additional Nuclear Technologies for Space Exploration

In 1961, the US Navy's Transit 4A navigation satellite became the first US spacecraft to be powered by nuclear energy. Transit 4A was powered by a Radioisotope Thermoelectric Generator (RTG), developed by the Atomic Energy Commission, the predecessor to the Energy Department. Since then, eight more generations of radioisotope power systems were developed by the Energy Department for use in space by NASA, the US Navy, and the US Air Force.

With no moving parts, RTGs convert heat from plutonium-238 decay into electricity using devices called thermocouples. The RTG on the Navy's Transit 4A satellite produced 2.7 W of electrical power. Transit 4A held the record for oldest broadcasting spacecraft for its first decade in orbit, during which time it traveled nearly 2 billion miles and circled the Earth more than 55,000 times.

In 1969, NASA launched the RTG-powered Nimbus III (i.e., Fig. 2.37), the first US weather satellite to measure air pressure, solar ultraviolet radiation, the ozone layer, and sea ice during both day and night.

Nimbus also included on-board infrared sensors that took early satellite photographs of the Earth. Aside from its RTGs, Nimbus also drew power from 10,500 built-in solar cells.

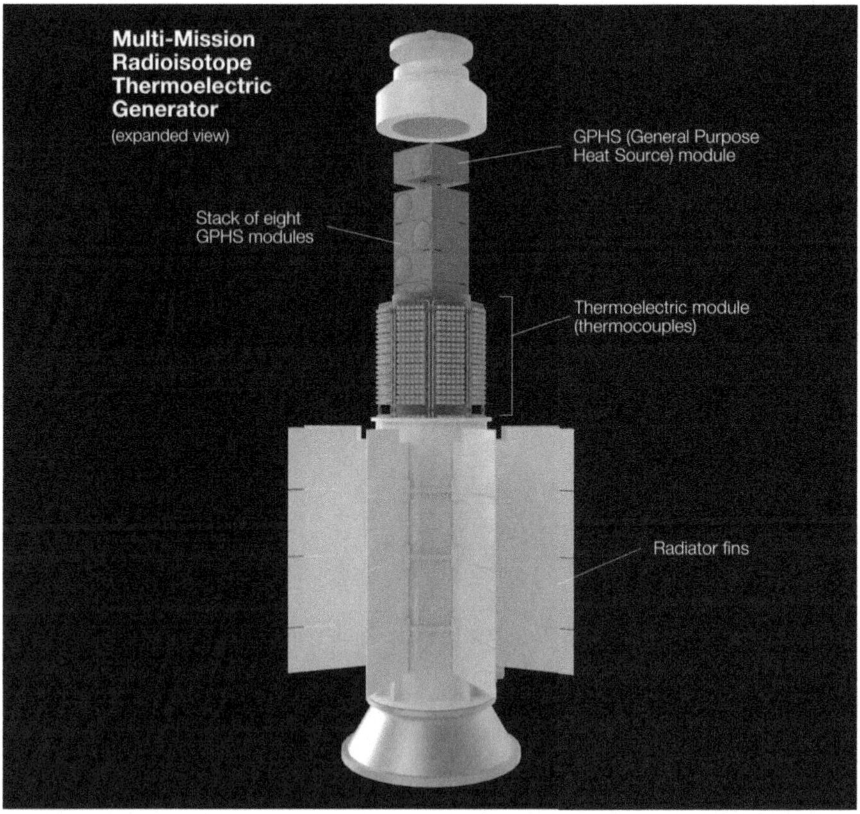

Fig. 2.36 Multi-Mission Radioisotope Thermoelectric Generator (RTG). (Source: NASA)

The space batteries on the moon also was part of NASA Research & Development (R&D) effort by NASA.

The Apollo missions to the moon included experimental packages for Apollo Lunar Surface Experiment Package (ALSEP)—containing scientific instruments that were left on the moon by US astronauts to send data back to Earth. The first package was solar-powered but relied on two 15-W Radioisotope Heater Units (RHUs) to keep its instruments warm enough to function.

The subsequent packages were each powered by 70-W SNAP-27 Radioisotope Thermoelectric Generators (RTGs). The ALSEPs contributed to a significant amount of what we now know about the moon—including data on solar wind and radiation, and the observation that the moon is geologically active. The five ALSEP stations were shut down in 1977.

RTGs have also powered missions to explore other planets. In 1989, Galileo became the first spacecraft to orbit Jupiter. Galileo showed evidence of an ocean of liquid water on Europa, one of Jupiter's moons, and volcanoes on Io, another moon; and took the first close-up pictures of an asteroid and the first photos of a comet

Fig. 2.37 Nimbus III image. (Source: NASA)

colliding with a planet, when Shoemaker-Levy 9 struck Jupiter. The Galileo orbiter was powered by two RTGs and included 120 RHUs to ensure its scientific instruments functioned properly.

The Ulysses mission to study the heliosphere—the part of space that is affected by the Sun's magnetic field—launched in 1990, powered by a General Purpose Heat Source Radioisotope Thermoelectric Generator (GPHS-RTG). It operated for about two decades before being powered down, during which time the Ulysses spacecraft flew past Jupiter and made three full polar orbits of the sun. Ulysses gathered previously unknown data about solar storms, solar wind, interstellar dust particles, and cosmic radiation. It also discovered 30 times more dust coming into the solar system from deep space than scientists had originally expected.

Cassini, an ongoing international mission to explore Saturn and its moons, is powered by three RTGs and kept warm by 117 small, strategically placed RHUs—82 on the Cassini orbiter and 35 on the Huygens probe, which Cassini carried to and released over Saturn's moon Titan. On January 14, 2005, Huygens successfully landed on Titan's surface, the first-ever landing of a craft from Earth in the outer solar system. Cassini is also responsible for the first comprehensive study of the Saturn system from orbit—including discoveries of active, icy geysers on Enceladus, another moon. Data collected from the Cassini mission is helping scientists understand more about what the Earth may have been like before life evolved.

Launched in 2006, the New Horizons spacecraft was designed to study Pluto and to explore other little-known, ice-cold places in the Kuiper Belt. New Horizons passed Jupiter and captured photos of that planet's rings and lightning at its poles before completing its flyby of Pluto in July 2015, returning the highest resolution images ever captured of the dwarf planet and its moons. It later reached Kuiper Belt object Ultima Thule on January 1, 2019, completing the farthest planetary flyby ever in exploration history at more than 4 billion miles away from us, here on Earth. The spacecraft is powered by a GPHS-RTG, similar to the one used on Ulysses.

NASA and DOE have explored other types of nuclear power technology over the years, including space nuclear reactors and nuclear propulsion technologies. Continued research and development of these and other related technologies might one day enable space missions to deliver more payloads on cargo missions, achieve faster trip times on piloted missions, or even provide power for crew stations on the surface of Mars or the moon.

2.9 Reaching For Interstellar Space

Pioneer 10 and Pioneer 11, launched in the early 1970s (i.e., Fig. 2.38), were precursors to the Voyager missions that followed. The spacecraft were designed to travel far—each powered by four RTGs and kept warm by 12 RHUs—and to withstand intense radiation from planets further out in the solar system.

Pioneer 10's power systems were designed to last at least 5 years but operated for more than three decades before communications ceased. During that time, it was the first spacecraft to fly past Mars, visit (and photograph) Jupiter, cross the asteroid belt and transmit data about interplanetary particles.

Pioneer 11 took the first up-close pictures of Saturn, discovered two new moons and an additional ring around the planet, and found that Saturn emits more than twice as much heat as it receives from the Sun. The mission lasted 22 years before communications ceased; now, Pioneer 10 and 11 are headed toward the edge of the solar system, bearing plaques with a message for intelligent beings they may encounter.

Voyager 1 and 2 built on Pioneer's legacy in the late 1970s. Taken together, these two missions have yielded some of the most important discoveries in US space exploration history. Each spacecraft uses nine RHUs to stay warm and draws power from three multi-hundred watt radioisotope thermoelectric generators, or MHW-RTGs—a type of power system specific to these two missions. The power systems are still operating today, more than 35 years after they were deployed. As the Voyager spacecraft slowly lose power, mission controllers back on Earth may turn off instruments one by one to conserve energy as long as possible.

Voyager 1 flew by Jupiter and Saturn and recently entered interstellar space. Along the way, it photographed the Earth—appearing as a pale blue dot—from 4 billion miles away. Voyager 2 is the only spacecraft to study all four of the giant planets—Jupiter, Saturn, Uranus, and Neptune—at close range. It discovered previously

Fig. 2.38 An artist's impression of a Pioneer spacecraft. (Source: NASA)

unknown moons around Neptune and Uranus, as well as liquid nitrogen geysers on Triton, one of Neptune's moons, and transmitted back hundreds of never-before-seen images. See Fig. 2.39.

Voyager 2 image of Fig. 2.39 shows the Triton from the south polar region with dark streaks produced by geysers visible on the icy surface.

Astronomers using the Gemini Observatory explore Neptune's largest moon Triton and observe, for the first time beyond the lab, an extraordinary union between carbon monoxide and nitrogen ices. The discovery offers insights into how this volatile mixture can transport material across the moon's surface via geysers, trigger seasonal atmospheric changes, and provide a context for conditions on other distant, icy worlds.

2.10 Mission to Mars

Viking 1 and 2, launched separately in 1975, were NASA's first effort to harvest data directly from the surface of the red planet. Each mission had two parts: an orbiter and a lander. Both Viking missions sent back photographs of the surface of the red

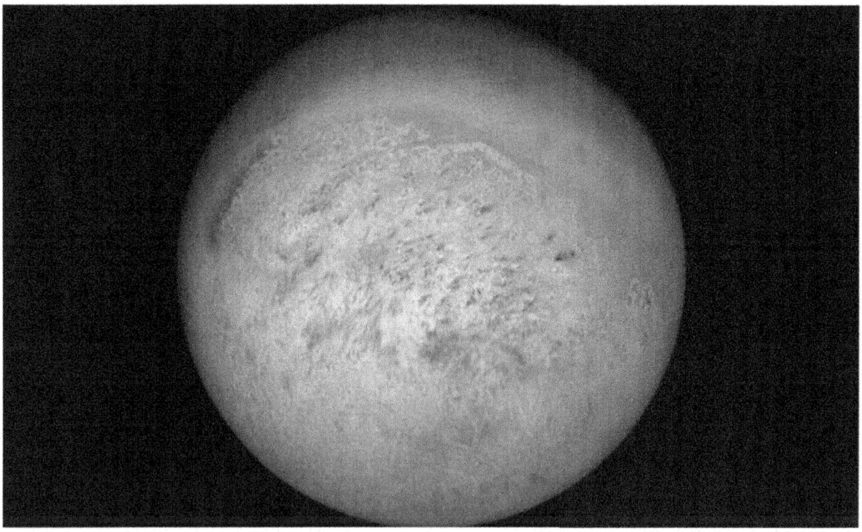

Fig. 2.39 Neptune's Moon Triton Fosters Rare Icy Union (Voyager 2 image). (Source: NASA/JPL)

planet and helped scientists back on Earth learn more about elements present there (carbon, nitrogen, hydrogen, oxygen, and phosphorus—all essential to life on our own home planet). The two 42.6-W RTGs on Viking 1 and 2 were designed to last at least 90 days but lasted for 6 and 4 years, respectively.

Interestingly, Viking 1 was not the first spacecraft to land on Mars—although it was the first successful one. A failed Soviet mission touched down on the Martian surface in 1971 but only survived for seconds before losing communication. Between Viking 1 and 2, more than 55,000 images of Mars were transmitted back to Earth—including the first space "selfie" on Mars, taken by Viking 2 itself. The image is one of the most famous pictures in the history of the US space program. See Fig. 2.40.

NASA took Mars exploration one step further in 1996, launching the microwave oven-sized Mars Pathfinder rover. Designed to last 7 days, the mission endured 12 times longer—demonstrating a cost-effective way to send a scientific mission to the red planet. Pathfinder used solar panels for electric power and relied on three RHUs to keep its scientific instruments warm.

In 2003, NASA separately launched twin rovers Spirit and Opportunity, designed to search Mars for evidence of water, climate changes, and other clues that the planet may have once supported life. Both rovers used solar panels for power and RHUs to support on-board scientific instruments. Spirit explored the red planet for 6 years—finding strong evidence that at one time Mars was much wetter than it is now—before it got stuck in sand and went no further. Opportunity has lived on, studying rock layers and returning stunning photographs of the Martian landscape.

The most recent nuclear-powered space mission to launch was 2011s Curiosity, which tweets from space. For both heat and power, Curiosity relies on a single multi-mission radioisotope thermoelectric generator that was constructed, assembled, and tested by the Energy Department and the Idaho, Oak Ridge, Los Alamos, and Sandia National Laboratories. This SUV-sized rover contains numerous scientific instruments and was sent to Mars to study rock layers and climate, determine if favorable conditions for life ever existed there, and to pave the way for future human exploration. Curiosity is far more powerful than its red planet predecessors and is expected to last for at least 2 years—presumably drilling and analyzing rock samples across an unprecedented range.

NASA plans to send a similar rover back to Mars in 2020. The rover is based on Curiosity's design, but this time will carry a drill for coring samples from Martian rocks and soil. The mission is expected to launch in July 2020 and land on Mars in February 2021. The RTG is currently being constructed, assembled, and tested by Idaho, Oak Ridge, Los Alamos, and Sandia National Laboratories.

Fig. 2.40 Artist's impression of the Viking Orbiter Spacecraft. (Source: NASA)

2.11 NASA Kilopower Reactor-Driven Future Space Exploration

The Kilopower project is a near-term technology effort to develop preliminary concepts and technologies that could be used for an affordable fission nuclear power system to enable long-duration stays on planetary surfaces. After successful completion of the Kilopower Reactor Using Stirling Technology (KRUSTY) experiment in March 2018, the Kilopower project team is developing mission concepts and performing additional risk reduction activities to prepare for a possible future flight demonstration. Such a demonstration could pave the way for future Kilopower systems that power human outposts on the Moon and Mars, enabling mission operations in harsh environments and missions that rely on In situ Resource Utilization to produce local propellants and other materials.

The Kilopower project is part of NASA's Space Technology Mission Directorate's Game Changing Development (GCD) program, which is managed by NASA's Langley Research Center. The Flight Opportunities program funded the parabolic and suborbital flights that matured the Kilopower technology's titanium water heat pipes by exposing it to space-relevant environments through the use of commercial reusable suborbital launch vehicles. The project will remain a part of the GCD program with the goal of transitioning to the Technology Demonstration Mission program in Fiscal Year 2020.

The Kilopower project team is led by NASA's Glenn Research Center in partnership with NASA's Marshall Space Flight Center; the Department of Energy's (DOE) National Nuclear Security Administration (NNSA); and NNSA's Los Alamos National Laboratory, Y-12 National Security Complex, and Nevada National Security Site (NNSS). Testing occurred from November 2017 to March 2018 at the National Criticality Experiments Research Center in the Device Assembly Facility at the NNSS.

Overall future of NASA space exploration depicted in Fig. 2.41.

Picture of NASA Kilopower reactor 1–3 kWe Multi-Mission is depicted in Fig. 2.42.

As it can be seen from Fig. 2.42, the NASA Kilopower reactor is using Advanced Sodium Hear Pipe Technology for reactor coolant [2, 3].

Throughout 2015, several material tests were initiated to understand certain properties that were either unavailable or were considered to be inconclusive based on past research data. Some of these tests included creep properties of the fuel, coefficient of thermal expansion of the fuel, and diffusion properties between the fuel and sodium heat pipes.

The development of NASA's Kilopower fission reactor is taking large strides toward flight development with several successful tests completed during its technology demonstration trials. The Kilopower reactors are designed to provide 1–10 kW of electrical power to a spacecraft or lander, which could be used for additional science instruments, the ability to power electric propulsion systems, or support human exploration on another planet. Power-rich nuclear missions have been excluded from NASA mission proposals because of the lack of radioisotope fuel and the absence of a flight qualified fission system. NASA has partnered with the Department of Energy's National Nuclear Security Administration to develop the Kilopower reactor using existing facilities and infrastructure and determine if the reactor design is suitable for flight development. The 3-year Kilopower project started in 2015 with a challenging goal of building and testing a full-scale flight-prototypic nuclear reactor by the end of 2017.

Initially, the power system will undergo several non-nuclear tests using an electrical heat source and a depleted uranium core to verify the complete non-nuclear system design prior to any nuclear testing. After successful completion of the depleted uranium test, the system will be shipped to the Nevada National Security Site where it will be fueled with the highly enriched uranium core and retested using the nuclear heat source. At completion of the project, NASA will have a significant sum of experimental data with a flight-prototypic fission power system, greatly reducing the technical and programmatic risks associated with further flight development. To complement the hardware-rich development progress, a review of several higher power mission studies are included to emphasize the impact of having a flight qualified fission reactor. The studies cover several science missions that offer nuclear electric propulsion with the reactor supplying power to the spacecraft's propulsion system and the science instruments, enabling a new class of outer planet missions. A solar versus nuclear trade for Mars surface power is also reviewed to compare the advantages of each system in support of ascent vehicle propellant production and human expeditions.

These mission studies offer insight into some of the benefits that fission power has to offer but still lacks a wider audience of influence. For example, mission directorates will not include a fission power system in their solicitations until it is flight

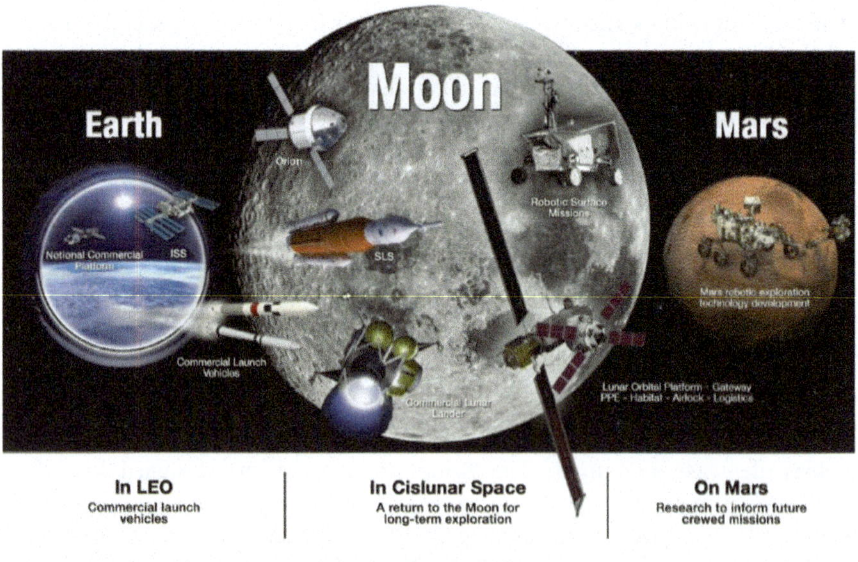

Fig. 2.41 Future NASA space exploration. (Source: NASA)

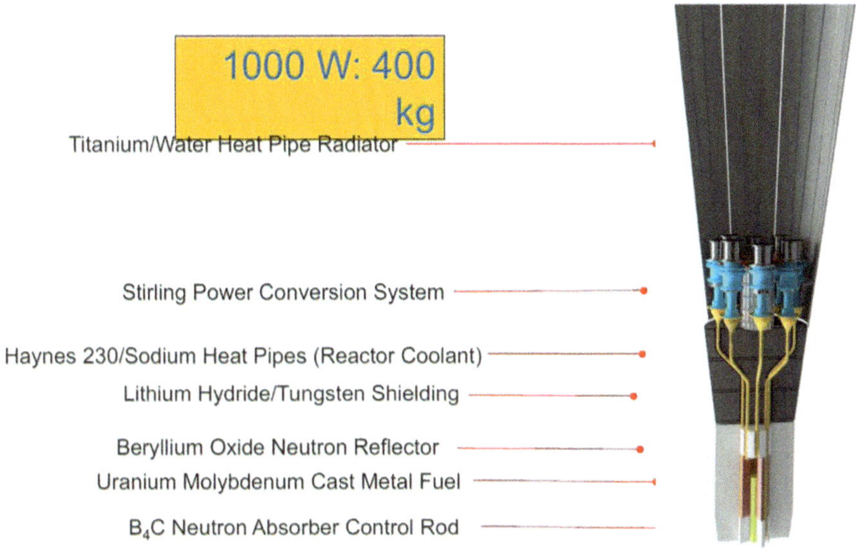

Fig. 2.42 NASA Kilopower reactor. (Source: NASA)

qualified, and scientists will not propose new missions that require more power than what is currently proven and available. An attempt to break this chicken and egg effect has been ongoing with the Kilopower project with the goal of advancing the technology to a level that encourages a flight development program and allows scientists to propose new ideas for higher power missions.

2.11.1 NASA Kilopower and What Is Next?

When astronauts someday venture to the Moon, Mars, and other destinations, one of the first and most important resources they will need is power. A reliable and efficient power system will be essential for day-to-day necessities, such as lighting, water, and oxygen, and for mission objectives, like running experiments and producing fuel for the long journey home.

That is why NASA is conducting experiments on Kilopower, a new power source that could provide safe, efficient, and plentiful energy for future robotic and human space exploration missions.

This pioneering space fission power system could provide up to 10 kW of electrical power—enough to run several average households—continuously for at least 10 years. Four Kilopower units would provide enough power to establish an outpost.

The prototype power system was designed and developed by NASA's Glenn Research Center in collaboration with NASA's Marshall Space Flight Center and the Los Alamos National Laboratory, while the reactor core was provided by the

Y12 National Security Complex. NASA Glenn shipped the prototype power system from Cleveland to the Nevada National Security Site (NNSS) in late September. See Fig. 2.43.

The team at the NNSS recently began tests on the reactor core. According to NASA Glenn's Marc Gibson, the Kilopower lead engineer, the team will connect the power system to the core and begin end-to-end checkouts this month. Gibson says the experiments should conclude with a full-power test lasting approximately 28 h in late March.

The advantages of Kilopower is explained as follows:

Fission power can provide abundant energy anywhere we want humans or robots to go. On Mars, the sun's power varies widely throughout the seasons, and periodic dust storms can last for months. On the Moon, the cold lunar night lingers for 14 days.

"We want a power source that can handle extreme environments," says Lee Mason, NASA's principal technologist for power and energy storage. "Kilopower" opens up the full surface of Mars, including the northern latitudes where water may reside. On the Moon, Kilopower could be deployed to help search for resources in permanently shadowed craters.

In these challenging environments, power generation from sunlight is difficult and fuel supply is limited. Kilopower is lightweight, reliable, and efficient, which makes it just right for the job.

Fig. 2.43 The Kilopower reactor. (Source: NASA)

2.12 Canada Driving Modular Nuclear Micro Reactor

As we stated in Chap. 1 of this book Canada is also pushing its demand for energy toward small modular and micro nuclear, and their attention toward micro reactor by reviewing their pre-licensing and demoing it by 2025 for a 4 MW size reactor that they are calling U-Battery type as its infrastructure layout shows in Fig. 2.44.

Canada in March of 2017 start U-Battery pre-licensing process according to the report by World Nuclear News (WNN) in 3 March 2017 [18].

The U-Battery consortium, led by Urenco, has registered its micro-modular reactor technology for pre-licensing vendor design review with the Canadian Nuclear Safety Commission (CNSC).

U-Battery is a "micro" nuclear reactor which will be able to produce local power and heat for a range of energy needs, mainly targeting the markets for industrial power units and off-grid locations. Powered by TRISO fuel, each helium gas-cooled unit produces 10 MWt, can deliver up to 4 MWe as electricity and can provide 750 °C of process heat. TRISO fuel comprises spherical particles of uranium fuel with a triple carbon-coating which effectively gives each tiny particle its own primary containment system. See Fig. 2.45.

The concept design of U-Battery was developed by the Universities of Manchester, the Dalton Institute (UK) and the Technology University of Delft (Netherlands) after the project was initiated in 2008 by Urenco. It is being developed by a consortium of Amec Foster Wheeler, Cammell-Laird, Laing O'Rourke, and Urenco.

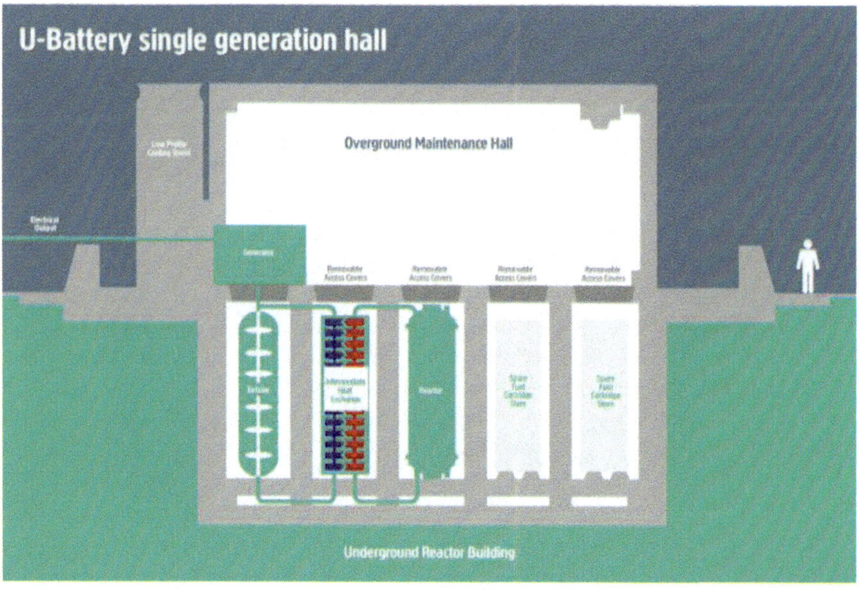

Fig. 2.44 4 Megawatt modular micro reactor infrastructure layout facility

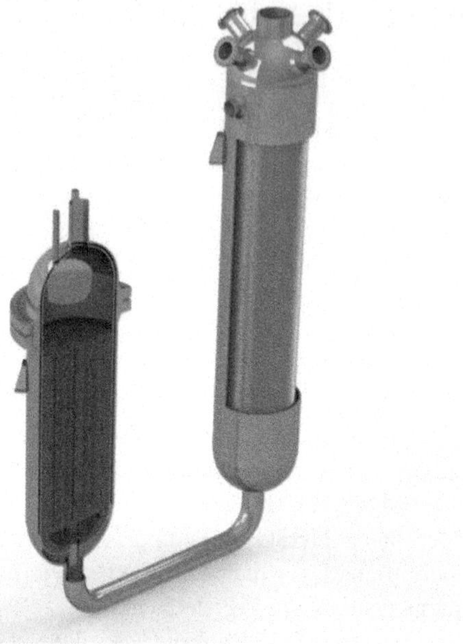

Fig. 2.45 Micro reactor modular image

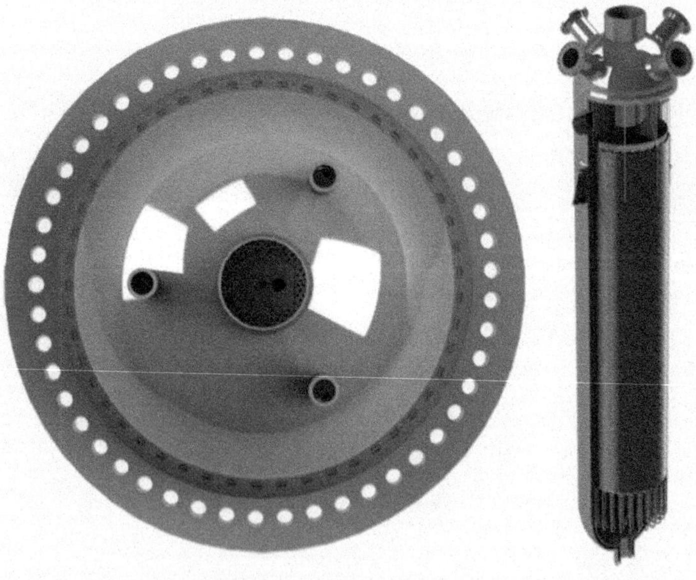

Fig. 2.46 TRISO fuel construction

The consortium aims to have a demonstration reactor operating by 2025 and estimates that by the fourth-of-a-kind unit, U-Battery's capital costs will be between £40 and £70 million ($49 and $86 million).

The core technical specifications of U-Battery are illustrated in Fig. 2.46:

- Twin unit—each unit has an output of 4 MW electric, 10 MW thermal.
- Gas cooled—helium in primary circuit, nitrogen in secondary circuit (no water).
- TRISO fuel—high integrity. In combination with low absolute power and absence of water eliminates need for multiple backup safety systems.
- Heat and power source—710 °C process heat.

As Fig. 2.33 shows, TRISO fuel is constructed by triple coating spherical particles of uranium fuel. A uranium center is coated in a layer of carbon, which in turn is coated in silicon carbide, with a further outer layer of carbon.

The structure and spherical shape of TRISO fuel means that it maintains its integrity under extreme conditions. Research has shown that even at temperatures of up to 1800 °C (200 °C hotter than assumed accident conditions) the majority of fission products remain inside the TRISO fuel particles, significantly enhancing safety.

TRISO fuel is based on proven technology. It was originally developed in the 1980s and is currently being manufactured in the USA. See Fig. 2.47.

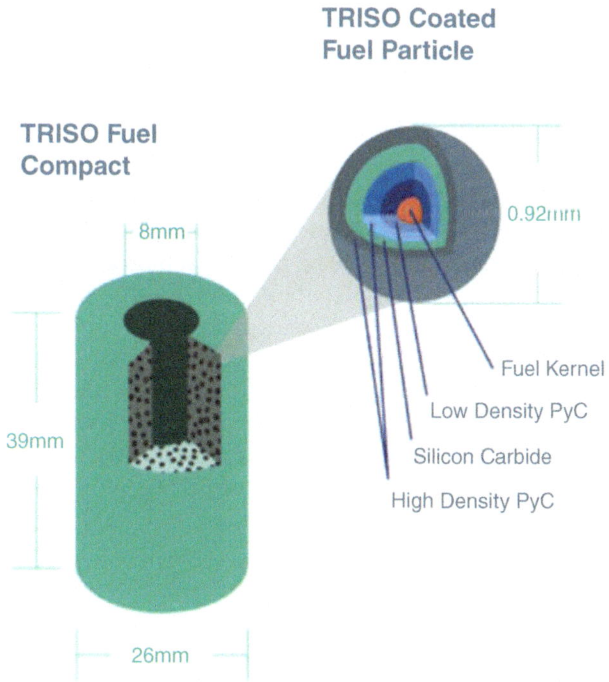

Fig. 2.47 TRISO fuel layout

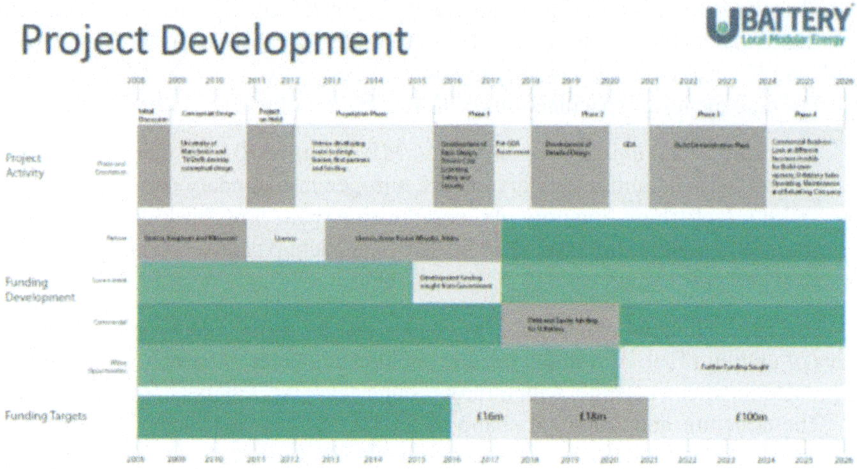

Fig. 2.48 Micro reactor project plan development in Canada

The project plan for implementation of this micro reactor from concept to production is depicted in Fig. 2.48.

Speaking at the UK Nuclear Industry Association's conference on small modular reactors (SMRs) earlier this week, Threlfall said the Canadian market for U-Battery could be "very, very large," with more than 300 locations, each of which could use between one and six batteries.

Threlfall said the consortium would be looking at when it could "get involved" in licensing in the UK. Triso fuel is already proven and is manufactured in the USA by BWXT, but Threlfall said most of the reactor's components could be supplied from the UK.

The CNSC's pre-licensing vendor review process is an optional service to provide an assessment of a nuclear power plant design based on a vendor's reactor technology. The three-phase review is not a required part of the licensing process for a new nuclear power plant but aims to verify the acceptability of a design with respect to Canadian nuclear regulatory requirements and expectations. Last year, the Canadian regulator agreed to conduct a first-phase vendor design review for LeadCold Reactor Inc's SEALER design concept and Terrestrial Energy's integral molten salt reactor design concept. In addition, Canadian reactor designer StarCore Nuclear applied in November to begin the vendor design review process for its 20 MWe high temperature gas reactor, which like U-Battery uses Triso fuel.

References

1. https://www.world-nuclear.org/information-library/nuclear-fuel-cycle/nuclear-power-reactors/small-nuclear-power-reactors.aspx
2. B. Zohuri, *Heat Pipe Application in Fission Driven Nuclear Power Plants*, 1st edn. (Springer Publishing Company, Cham, 2019)
3. B. Zohuri, *Heat Pipe Design and Technology: Modern Applications for Practical Thermal Management*, 2nd edn. (Springer Publishing Company, Cham, 2016)
4. https://www.powermag.com/big-gains-for-tiny-nuclear-reactors/
5. B. Zohuri, P. McDaniel, *Advanced Smaller Modular Reactors: An Innovative Approach to Nuclear Power*, 1st edn. (Springer, Cham, 2019). https://www.springer.com/us/book/9783030236816
6. B. Zohuri, P. McDaniel, *Combined Cycle Driven Efficiency for Next Generation Nuclear Power Plants: An Innovative Design Approach*, 2nd edn. (Springer, Cham, 2018). https://www.springer.com/gp/book/9783319705507
7. B. Zohuri, P. McDaniel, C.R. De Oliveria, Advanced nuclear open air-Brayton cycles for highly efficient power conversion. Nucl. Technol. **192**(1), 48–60 (2015). https://doi.org/10.13182/NT14-42
8. B. Zohuri, *Combined Cycle Driven Efficiency for Next Generation Nuclear Power Plants: An Innovative Design Approach* (2016)
9. C. Yu, Small Nuclear Reactors, Physics 241, Stanford University, Winter 2011
10. Small Modular Reactors: A Window on Nuclear Energy, Andlinger Center, Princeton University, June 2015
11. S. Harber, Small Nuclear Reactors: Background, Potential Applications, and Challenges, 19 Feb 2017. http://large.stanford.edu/courses/2017/ph241/harber1/
12. K. Stacey, Small Modular Reactors Are Nuclear Energy's Future, *Financial Times*, 25 July 2016
13. https://www.nei.org/resources/reports-briefs/road-map-micro-reactors-defense-department
14. https://www.acq.osd.mil/dsb/reports/2000s/ADA477619.pdf
15. DoD Annual Energy Management and Resilience Report FY2016
16. https://www.rand.org/content/dam/rand/pubs/research_reports/RR2000/RR2066/RAND_RR2066.pdf
17. DoD Summary of the National Defense Strategy of the United States of America
18. http://www.world-nuclear-news.org/RS-U-Battery-begins-Canadian-pre-licensing-process-0303177.html

Chapter 3
Nuclear Micro Reactor Research, Development, and Deployment

3.1 Introduction

The US Department of Energy (DOE) has opened the National Reactor Innovation Centre (NRIC) to support the development of advanced nuclear energy technology.

A result of the Nuclear Energy Innovation Capabilities Act (NEICA), the new center is located at the Idaho National Laboratory (INL) in Idaho Falls, USA.

The act was signed into law in 2018 to accelerate the development of advanced reactors in the country by eliminating financial and technical barriers.

The US energy secretary said: "NRIC will enable the demonstration and deployment of advanced reactors that will define the future of nuclear energy."

He also stated, "By bringing the industry together with our national labs and university partners, we can enhance our energy independence and position the US as a global leader in advanced nuclear innovation."

Technology developers from the private industry will receive support from NRIC required to test, demonstrate, and evaluate the performance of their reactor concepts.

It will also help technology developers in speeding-up the licensing and commercialization of their new nuclear energy systems. As per Idaho Senator Mike Crapo, "This is the ultimate public–private sector partnership that will take the next step to grow clean nuclear power across the nation."

"It makes sense that the INL is the place where private developers will now work with the Department of Defense and Nasa to come and demonstrate new nuclear technologies. I look forward to seeing micro reactors and small modular reactors being built here. The best is yet to come!"

NRIC aims to demonstrate small modular reactor and micro reactor concepts in the next 5 years.

In the fiscal year 2020 budget, the House Energy and Water Development committee had set aside $5 million for NRIC [1].

B. Zohuri, *Nuclear Micro Reactors*, https://doi.org/10.1007/978-3-030-47225-2_3

"If you put yourself in the shoes of an investor-owned utility and they're looking at a combined cycle gas plant that they could build in 18 months for an installed kilowatt hour cost of 700 or 800 bucks, versus a nuclear project that might take 10 years or more and cost 10 billion or more... That's a very hard sell," Daniel Poneman, former US Deputy Secretary of Energy, and current president and CEO of nuclear developer Centrus, told Utility Dive Tuesday at the 2019 New Nuclear Capital Conference in Washington, DC.

Investor and owner of utility companies around the world, especially in the USA are looking at a combined cycle gas power plant presently that within period of 18 build of such plant after installation they can produce electricity at cost of \$700–\$800 per kWh, versus a nuclear power plant project for traditional type nuclear reactor of Generation III (Gen-III) that might take 10 years or more at cost of nearly \$10 billion dollars or more (Fig. 3.1).

As we stated at the beginning of this chapter, the large-scale reactors that are the norm across the USA do not fit in with the growing trend toward smaller, decentralized power, when the demand for generating electricity exponentially growing as a function population growth approximately at the rate of 17% per year, globally.

With new innovative combined cycle approach, thermodynamically and increasing thermal efficiency output of new generation of small and advanced nuclear reactor, the cost of ownership of these type of power plant decreasing [2–7].

With this innovative approach, small modular and micro reactors of advanced types (i.e., GEN-IV and Nuclear Miro Reactor) will better fit the above mold, but yet to enter the market and change adverse public opinion that exists to a proponent form, given the recent disaster and accident at Fukushima Daiichi nuclear site (2011) in Japan.

There is no question in anybody's mind that the demand for electricity globally and annually is on the rise, and economically the renewable source of energy using wind and solar along will not cut it alone, thus need for nuclear energy in an invadable circumstances.

Fig. 3.1 A traditional nuclear power plant facility

Thus, the key question, say stakeholders, is how to spur that initial investment and establish a commercial domestic market, with a loftier goal of establishing the USA as a nuclear power export leader.

The key issue that impact the public opinion is safety operational assurance of these generation of reactor that we have discussed in Sect. 3.2 of this chapter.

Recent congress direction to Department of Energy (DOE) to continue with study of Micro Reactor Deployment the need to satisfy the Department of Defense (DOD) for such technology for its mobility deployment of its brigade globally is on the rise. Industries such as NuScale, Holtic, and Hitachi/GE with their approach to Advanced Small Modular Reactor and Westinghouse with its eVinci Advanced Heat Pipe [8, 9] Nuclear Micro Rector have gain momentums in the past decade. See Chap. 2 of this book as well.

- DOE study would list possible military installations for micro reactors
- Study would include timeline, cost-estimates for micro reactor pilot program
- Bill also lets DOE secretary delegate many export control approvals

The National Defense Authorization Act (NDAA) passed by Congress this week includes language directing the Secretary of Energy to develop a report on a pilot program for deploying micro reactors at national security facilities.

The Department of Energy (DOE) report identifies where micro reactors could play a vital role in helping to harden and secure power generation at Department of Defense (DOD) and DOE facilities.

"As an emphasis continues to be placed on clean energy, countries around the globe are pursuing nuclear power for the carbon-free reliable electricity it produces," Nuclear Energy Institute (NEI) President and CEO Maria Korsnick said.

"This bill helps to ensure American companies are at the table with our foreign competitors so that we can continue to provide our technology and, in doing so, set international nonproliferation, security and safety standards for years to come."

The bill defines micro reactors as reactors with a "capacity that is not greater than 50 MW." These advanced reactor technologies have not submitted license applications to the Nuclear Regulatory Commission and could be deployed by the mid-2020s.

The legislation directs DOE to develop a report to:

- Identify potential locations to site, construct, and operate a micro reactor on DOD or DOE facilities that contain critical national security infrastructure.
- Assess different nuclear technologies to provide energy resiliency for critical national security infrastructure.
- Survey potential commercial stakeholders with which to enter into a contract under the pilot program to construct and operate a licensed micro reactor.
- Estimate costs of the pilot program.
- Provide a timeline of pilot program milestones.

The bill gives the DOE one year from the time of the bill's enactment to submit its report to Congress. The President is expected to sign the bill into law in the coming days.

This bill also allows DOE secretary to delegate export control decision-making as well.

The National Defense Authorization Act that Congress passed this week also includes an important provision allowing the DOE Secretary to delegate decision-making for commercial nuclear export controls to most countries, except with respect to enrichment and processing of special nuclear material.

Requiring the Secretary to approve every single Part 810 specific authorization adds substantial time to an already slow process. A recent letter [10] from experts in the field notes: "Permitting delegation of actions under Part 810 would allow the Secretary of Energy to focus his or her time on higher priority issues, would be a better use of taxpayer resources and would significantly reduce regulatory burden on the U.S. industry, without sacrificing national security interests." See Fig. 3.2.

According to a December 2017 report by the Nuclear Innovation Alliance (NIA) [11], DOE approvals of Part 810 applications often take more than a year on average. By comparison, other leading nuclear supplier countries require from 5 weeks to 3 months for an equivalent export authorization.

According to the report by NIA, the US Energy Information Administration (EIA) projects that by 2050 countries around the world will add almost 200 GW of new nuclear energy capacity. Those construction projects will entail the flow of new nuclear materials, services, and equipment to a number of countries that currently do not possess significant nuclear power programs. A growth in nuclear energy use offers major commercial opportunities for nuclear reactor companies and carries implications for the global nonproliferation regime. As Table 3.1 shows, most of the expected deployments are projected to take place in countries that are not members of the Organization for Economic Co-operation and Development (OECD).

Before the first reactors are under construction, however, supplier nations typically share proprietary information on their reactor designs with potential customer nations.

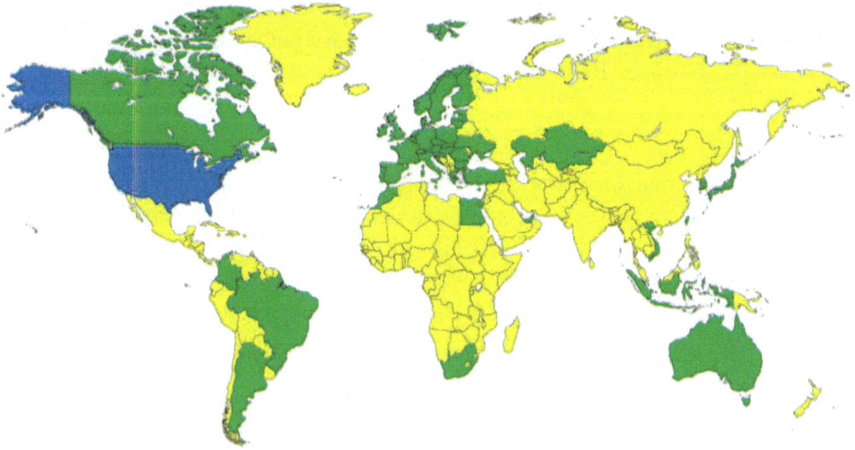

Fig. 3.2 Improving the efficiency of US export controls for nuclear energy technologies (Note: Green Zones are clear to obtain U.S. Nuclear Technology)

Table 3.1 EIA projection for additional nuclear energy capacity by region (capacity in gigawatts) (Source: Courtesy of EIA)

Region	2015	2030	2050	Change from 2015 to 2050
OECD Courtiers	256	259	200	−56
Non-OECD Europe and Eurasia	42	57	56	+14
Non-OECD Asia	39	124	231	+192
Non-OECD Americas	4	6	5	+1
Africa	2	4	6	+4
Middle East	1	12	17	+16
Total	**343**	**462**	**516**	**+173**

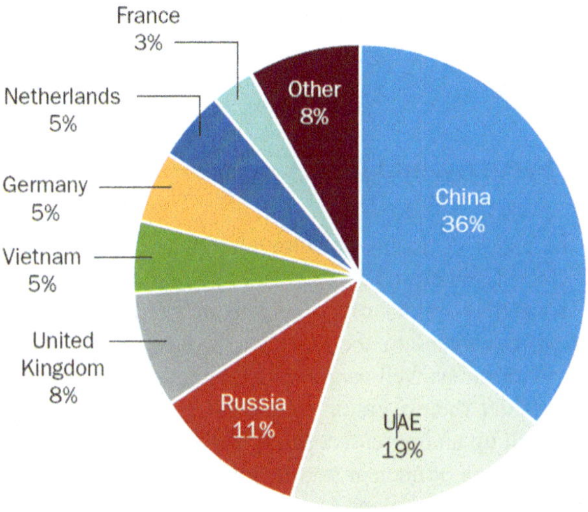

Fig. 3.3 Specific authorization by country, 2000–2015. (Source: DOE Reading Room) [11]

These transactions may be the first technology transfers where the government of a supplier nation will have to consider the commercial and nonproliferation implications of broader nuclear energy cooperation with a first-time nuclear energy customer nation. Even between countries where nuclear trade has been ongoing for decades, new transactions such as these may pose unique and complex challenges.

Specific authorizations by country are shown in Fig. 3.3. The 2000–2015 time period is used to give a more representative depiction of where specific authorization applications are targeting

though some changes have already taken place. For example, the UAE became a generally authorized destination in 2015, so activities there will no longer require permission from the US government (unless they involve activities noted in 810.7c).

As Table 3.2 shows, the application processing times can vary greatly depending on the destination country. This is partly a reflection of the response time of the destination country in providing the requested assurances but may also be driven by US considerations of the particular country involved.

Table 3.2 Specific authorization to Russia, China, and the UAE, 2007–2015 (Source: Data supplied by DOE/NNSA in response to an FOIA request by the Clean Air Task Force)

Country	Number of authorization	Average processing time (in days)
Russia	6	505
China	18	487
USA	12	193

In the USA, this intersection of business and national security takes place under the US Department of Energy (DOE) 10 CFR Part 810 (Part 810) regulations, which control the flow of unclassified nuclear energy technology and assistance to foreign atomic energy activities.

These regulations and their implementation are the subject of Nuclear Innovation Alliance report. [11]

3.2 Safety, Security, and Cost Concerns

Aftermath of the major accidents at Three Mile Island in 1979 and Chernobyl in 1986 and then, recent devastated Japan's Fukushima nuclear power plant frailer in Japan in March of 2011, pretty much nuclear power fell out of favor, and some countries applied the brakes to their nuclear programs. Concerns about climate change and air pollution, as well as growing demand for electricity, led many governments to reconsider their aversion to nuclear power, which emits little carbon dioxide and had built up an impressive safety and reliability record. Some countries reversed their phase-outs of nuclear power, some extended the lifetimes of existing reactors, and many developed plans for new ones.

Despite all these given concerns and issues in respect to the nuclear energy, still we are facing the fact of why we still need nuclear power as clean source of energy, particularly when, we deal with renewable source of energy arguments [12].

Today, roughly 60 nuclear plants are under construction worldwide, which will add about 60,000 MW of generating capacity—equivalent to a sixth of the world's current nuclear power capacity; however, this movement has been lost after March of 2001 and Japan's Fukushima nuclear power episode.

Nuclear power's track record of providing clean and reliable electricity compares favorably with other energy sources. Low natural gas prices, mostly the result of newly accessible shale gas, have brightened the prospects that efficient gas-burning power plants could cut emissions of carbon dioxide and other pollutants relatively quickly by displacing old, inefficient coal plants, but the historical volatility of natural gas prices has made utility companies wary of putting all their eggs in that basket. Besides, in the long run, burning natural gas would still release too much carbon dioxide. Wind and solar power are becoming increasingly widespread, but their intermittent and variable supply make them poorly suited for large-scale use in the absence of an affordable way to store electricity. Hydropower, meanwhile, has very limited prospects for expansion in the USA because of environmental concerns and the small number of potential sites [13].

As part of any nuclear power plant safety that one should consider as part of design and operation of such source of energy is the reactor stability. Understanding time-dependent behaviors of nuclear reactors and the methods of their control is essential to the operation and safety of nuclear power plants. This chapter provides researchers and engineers in nuclear engineering very general yet comprehensive information on the fundamental theory of nuclear reactor kinetics and control and the state-of-the-art practice in actual plants, as well as the idea of how to bridge the two. The dynamics and stability of engineering equipment that affects their economical and operation from safety and reliable operation point of view. In this chapter, we will talk about the existing knowledge that is today's practice for design of reactor power plants and their stabilities as well as available techniques to designers. Although stable power processes are never guaranteed. An assortment of unstable behaviors wrecks power apparatus, including mechanical vibration, malfunctioning control apparatus, unstable fluid flow, unstable boiling of liquids, or combinations thereof. Failures and weaknesses of safety management systems are the underlying causes of most accidents [14].

The safety and capital cost challenges involved with traditional nuclear power plants may be considerable, but a new class of reactors in the development stage holds promise for addressing them. These reactors, called small modular reactors (SMRs), produce anywhere from 10 to 300 MW, rather than the 1000 MW produced by a typical reactor. An entire reactor, or at least most of it, can be built in a factory and shipped to a site for assembly, where several reactors can be installed together to compose a larger nuclear power station. SMRs have attractive safety features, too. Their design often incorporates natural cooling features that can continue to function in the absence of external power, and the underground placement of the reactors and the spent-fuel storage pools is more secure.

Since Small Modular Reactors (SMRs) are smaller than conventional nuclear plants, the construction costs for individual projects are more manageable, and thus the financing terms may be more favorable. And because they are factory-assembled, the on-site construction time is shorter. The utility company can build up its nuclear power capacity step by step, adding additional reactors as needed, which means that it can generate revenue from electricity sales sooner. This helps not only the plant owner but also customers, who are increasingly being asked to pay higher rates today to fund tomorrow's plants [13].

With the US federal budget under tremendous pressure, it is hard to imagine taxpayers funding demonstrations of a new nuclear technology. But if the USA takes a hiatus from creating new clean-energy options—be it SMRs, renewable energy, advanced batteries, or carbon capture and sequestration—Americans will look back in 10 years with regret. There will be fewer economically viable options for meeting the energy and environmental needs of the USA, and the country will be less competitive in the global technology market.

Small modular reactors are unlikely to solve the economic and safety problems faced by nuclear power. According to the US Department of Energy (DOE) and some members of the nuclear industry, the next big thing in nuclear energy will be a small thing: the "small modular reactor" (SMR).

SMRs—"small" because they generate a maximum of about 30% as much power as typical current reactors, and "modular" because they can be assembled in factories and shipped to power plant sites—have been getting a lot of positive attention recently, as the nuclear power industry has struggled to remain economically viable in an era of flat demand and increasing competition from natural gas and other energy alternatives.

SMRs have been touted as both safer and more cost-effective than older, larger nuclear reactor designs. Proponents have even suggested that SMRs are so safe that some current NRC regulations can be relaxed for them, arguing that they need fewer operators and safety officers, less robust containment structures, and less elaborate evacuation plans. Are these claims justified?

3.2.1 Are Small Modular and Micro Reactors Safer?

One of the chief selling points for SMRs is that they are supposed to be safer than current reactor designs. However, their safety advantages are not as straightforward as some proponents suggest:

- SMRs and Micro Reactors use passive cooling systems that do not depend on the availability of electric power. This would be a genuine advantage under many accident scenarios, but not all. Passive systems are not infallible, and credible designs should include reliable active backup cooling systems. But this would add to cost.
- SMRs and Micro Reactors feature smaller, less robust containment systems than current reactors. This can have negative safety consequences, including a greater probability of damage from hydrogen explosions. SMR designs include measures to prevent hydrogen from reaching explosive concentrations, but they are not as reliable as a more robust containment—which, again, would add to cost.
- Some proponents have suggested siting SMRs underground as a safety measure. However, underground siting is a double-edged sword—it reduces risk in some situations (such as earthquake) and increases it in others (such as flooding). It can also make emergency intervention more difficult and it too increases cost.
- Proponents also point out that smaller reactors are inherently less dangerous than larger ones. While this is true, it is misleading because small reactors generate less power than large ones, and therefore more of them are required to meet the same energy needs. Multiple SMRs and Micro Reactors may actually present a higher risk than a single large reactor, especially if plant owners try to cut costs by reducing support staff or safety equipment per reactor.

3.2.2 *Shrinking Evacuation Zones*

Because of SMRs alleged safety advantages, proponents have called for shrinking the size of the emergency planning zone (EPZ) surrounding an SMR plant from the current standard of 10 miles to as little as 1000 ft, making it easier to site the plants near population centers and in convenient locations such as former coal plants and military bases.

However, the lessons of Fukushima, in which radiation levels high enough to trigger evacuation or long-term settlement were measured at as much as 20–30 miles from the accident, suggest that these proposals, which are based on assumptions and models that have yet to be tested in practice, may be overoptimistic.

3.3 Economies of Scale and Catch

SMR and Micro Reactors based power plants can be built with a smaller capital investment than plants based on larger reactors. Proponents suggest that this will remove financial barriers that have slowed the growth of nuclear power in recent years.

However, there is a catch: "affordable" does not necessarily mean "cost-effective." Economies of scale dictate that, all other things being equal, larger reactors will generate cheaper power. SMR proponents suggest that mass production of modular reactors could offset economies of scale, but a 2011 study [15] concluded that SMRs would still be more expensive than current reactors.

Even if SMRs could eventually be more cost-effective than larger reactors due to mass production, this advantage will only come into play when many SMRs are in operation. But utilities are unlikely to invest in Small Modular Reactors (SMRs) and Miro Reactors (MRs) until they can produce competitively priced electric power. This Catch-22 has led some observers to conclude that the technology will require significant government financial help to get off the ground. See Fig. 3.4.

UxC, LLC (UxC), a leader in global nuclear market analysis, is pleased to announce a new report on the *SMR Market Outlook* (SMO). This 450 page report provides comprehensive commercial and technical analysis of the overall SMR market and detailed reviews of the leading SMR designs in the world [15].

The SMO report is the latest result of UxC's continuing efforts to provide its global clients with insightful products and services in the SMR sector. Contents of the SMO include:

- Detailed commercial and technical analyses with "Pros vs. Cons" assessments of the world's top ten leading SMRs.
- Analysis of the key challenges for SMR deployment.
- Review of the SMR business case: economics and potential business models for deployment.
- Competitive analysis of SMR market: US vs. international vendors.

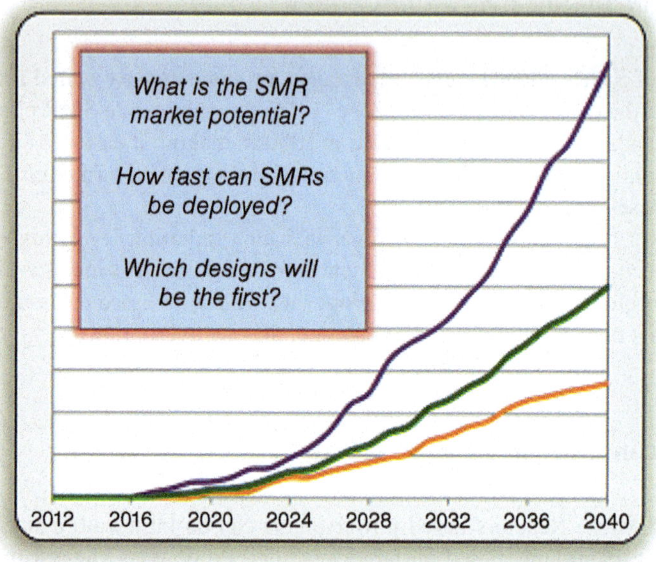

Fig. 3.4 SMRs market potentials [15]

- Profiles of SMR industries in major countries.
- Global outlook for SMR deployment: potential clients and their rationales.
- SMR deployment forecasts to 2040.

3.3.1 Building a Domestic Market

Many in the nuclear industry argue that establishing the USA as a nuclear export leader is essential to national security as well as global decarbonization. While the Pentagon can get things rolling for small-scale nuclear, utilities will be essential to building out the market as buyers.

"I think you see utilities, particularly in Canada, stepping up with the Canadian government to take on that market-making role and we look forward to similar discussions here in the U.S.," Donald Wolf, president and CEO of Advanced Reactor Concepts, a US-based developer, said at the New Nuclear Capital Conference. "Before we, in effect, push these new designs on foreign countries, it really helps to say we built it here first. We've improved it at home, and it's safe for us" [16].

And some developers are moving to establish those markets for small reactors with municipal power systems, which make for more attractive first customers than IOUs, according to Colbert. Municipal power systems have access to a lower weighted cost of capital, around 3–4%, compared to IOUs approximately 8–10%.

NuScale has jointly pursued both DOD and the Utah Associated Municipal Power Association (UAMP) as first customers. But the developer's first plant, a 12 module, 720 MW nuclear generator, is expected to be fully operational by 2027 under UAMP, according to Colbert. And the municipal utility on Wednesday announced it has sales contracts for enough power to begin licensing with the Nuclear Regulatory Commission.

"The challenge [with DOD contracting] has been for the Department of Defense to define what the requirements are, and when they want it, and how they would go about financing that," he said. "So far, that hasn't come together, but it seems like they're moving more toward that way."

A lot of interest in NuScale's reactors, he said, has come from international markets, including Romania, Jordan, and Canada, that want to see that the plant is operational before they invest.

"After that, the question is how quickly can I manufacture to scale," said Colbert. NuScale is in conversations with manufacturers BWX Technologies and Doosan Heavy Industries and Construction to ultimately start manufacturing "2–3 GW per year early on," with a wider aim of 35 GW by 2035.

3.4 Barrier to Nuclear

Advanced reactors are not just a garden science experiment, "said Murkowski." There is a real private sector interest. There is capital being invested to make [them] a reality, "but developing a first of its kind technology, we all know, is not for the risk averse, it's not for the faint of heart."

With large capital investments, such as nuclear, some of that risk can turn away private investment, said Colbert.

"If you look at nuclear, the market for energy … is pretty well known," he said. "What's not so well known is how do you get through the nuclear regulatory process, the [Nuclear Regulatory Commission] process."

While safety remains a paramount concern for nuclear power, many in the industry say they no longer view public perception as a major problem.

"How do we tell the masses the progress we're making? … How do we tell potential investors … that the risk of a nuclear event that contaminates and causes evacuation needs is gone for your technology?" an audience member asked panelists.

"I don't think that's the biggest challenge facing our industry," responded Clay Sell, CEO of nuclear engineer and design company X-energy. "The closer people look to a nuclear power plant and the more informed they are about nuclear power, the more accepting and understanding they are of the risk and the more they love it. And so that tells me that education wins."

3.5 Unrivaled Small Modular Reactors Credentials

UxC has been actively tracking the SMR sector for many years. Along the way, UxC's team of reactor market analysts has been collaborating with international consultants from industry and academia, who are authorities in reactor technology and design.

Our first achievement in SMR analysis was the publication in 2010 of UxC's *Small Modular Reactor Assessments* (SMRA) report. This groundbreaking, in-depth assessment of SMR technologies and the broader market was praised throughout the industry as a thoughtful, independent, and exhaustive study of the SMR landscape [15].

Moreover, UxC's experts have participated in the most important conferences, events, and initiatives related to SMRs over the years, including numerous invitations to deliver presentations on various topics of interest [15].

3.6 High-Assay Low Enriched Uranium (HALEU)

In this section, we cover the development of a US fuel source for advanced reactors very briefly.

When uranium is extracted from the earth, more than 99% of the atoms are U^{238}, while the fissionable isotope U^{235} represents less than 1%. To fuel one of today's commercial nuclear reactors, the uranium must be enriched so that the U^{235} concentration, or "assay," is raised between 4 and 5%. This is called Low Enriched Uranium (LEU).

Many of the reactor designs being prepared for future deployment will need a higher assay uranium fuel to operate. High-Assay Low Enriched Uranium (HALEU) has a U^{235} assay above 5% but below 20%. This is still far below the assay required to make weapons or to power US submarines and aircraft carriers (Fig. 3.5).

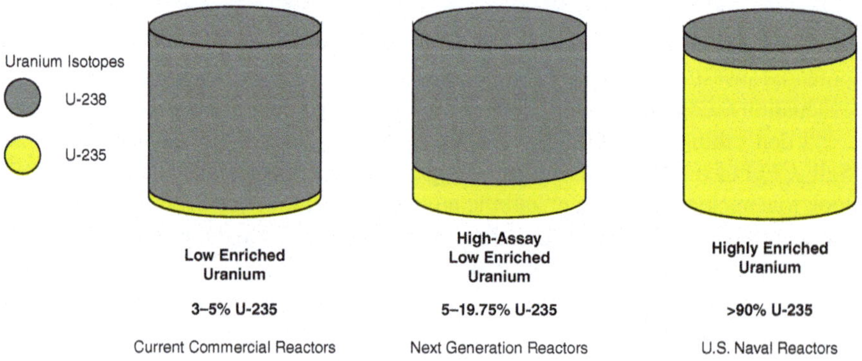

Fig. 3.5 High-assay low enriched uranium fuel level

HALEU fuel has many advantages that improve reactor performance. Because the U^{235} is more concentrated, the fuel assemblies and reactors can be smaller, which is one reason why many Small Modular Reactor (SMR) designs will run on HALEU. The reactors do not need to be refueled as often, and they can achieve higher "burnup" rates, which means less fuel will be required and less waste will be produced.

3.6.1 High-Assay Low Enriched Uranium (HALEU) Fuel Fabrication

Once uranium has been enriched to a level necessary for HALEU fuel, it will need to be fabricated into fuel forms for loading into a reactor. To that end, Centrus is working under a contract with X-energy, a pioneering reactor and fuel company, to pursue the development of a fuel fabrication facility that would produce X-energy's Uranium Oxycarbide (UCO) Tristructural Isotropic (TRISO) fuel forms using HALEU. This unique HALEU fuel could power a variety of advanced reactors under development around the world, ensuring a US source of supply for this emerging industry.

Estimated commercial HALEU market needs chart is depicted in Fig. 3.6.

Bear in mind the above chart does not project the account for potential DOD needs for their need for Nuclear Micro Reactors (NMRs).

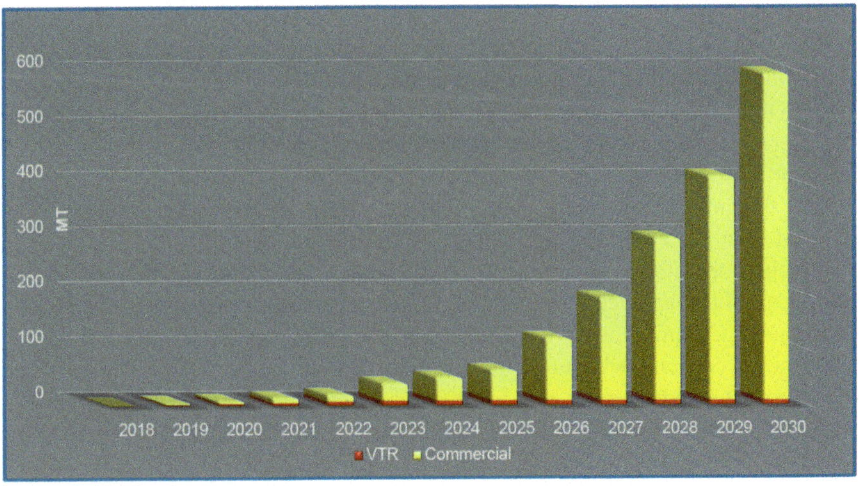

Fig. 3.6 Estimated commercial HALEU market needs. (Source for Industry Demand: Nuclear Energy Institute's Letter to Secretary of Energy, DOE, July 5, 2018)

3.6.2 US Enrichment Technology Demonstration

The company's such as Centrus is working with the US Department of Energy to deploy a small cascade of Centrus' AC100M centrifuges at the American Centrifuge Plant in Piketon, Ohio, to demonstrate the capability of our technology to produce HALEU fuel needed for advanced reactors.

3.7 Nuclear Power Pros and Cons

As we know based on physics of tradition Nuclear Power Reactors (NPP), nuclear power comes from the process of nuclear fission, or the splitting of atoms. The resulting controlled nuclear chain reaction creates heat, which is used to boil water, produce steam, and drive turbines that generate electricity.

Nuclear power comes from the process of nuclear fission, or the splitting of atoms. The resulting controlled nuclear chain reaction creates heat, which is used to boil water, produce steam, and drive turbines that generate electricity.

The USA is home to 99 nuclear power units, located in 31 states. Together, these plants generate almost 20% of America's electricity, or approximately 8.5% of its total energy as per DOE [17].

However, no new nuclear reactors have started up in the USA since 1996 [18].

The last newly built reactor to enter service was Watts Bar 1 in Tennessee, in 1996. In 2007, the Tennessee Valley Authority voted to complete construction of Watts Bar 2. On October 22, 2015, the Nuclear Regulatory Commission approved a 40-year operating license for Unit 2, marking the formal end of construction and allowing for the installation of nuclear fuel and subsequent testing. On December 15, 2015, TVA announced that the reactor was fully loaded with fuel and is ready for criticality and power ascension tests. Commercial Operation is expected in June 2016 [19].

From 1973 to 2010, electricity generated by nuclear plants in the USA rose TENFOLD to over 800,000 MWh, but output has declined slightly in recent years as several merchant plants have retired [20]. Reliability of existing plants has grown substantially during this time, which means that existing plants are producing more energy than in the past. Nuclear capacity factors are now averaging around 92%, up from 56% in 1975 (i.e., Fig. 3.7) [21].

Nuclear operators can increase the rated capacity of their plants through power up rates, which are license amendments that must be approved by the Nuclear Regulatory Commission (NRC). Up rates can vary from small (less than 2%) increases in capacity, which require very little capital investment or plant modifications, to extended up rates of 15–20%, requiring significant modifications. It is expected that 0.2 GW of additional nuclear power will be added between 2013 and 2040 through up rates at existing nuclear power plants [22]. The USA is the world's largest producer of nuclear power, but it derives a smaller percentage of its electricity from nuclear technology than many other industrial countries. In 2013, France derived 76% of its electricity from nuclear power. Other countries producing a high

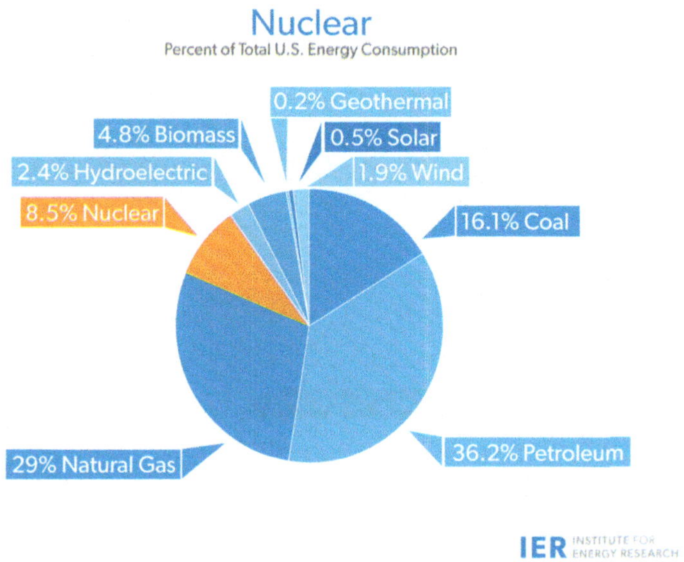

Fig. 3.7 Percent of total US energy consumption. (Source: EIA, MER, March 2016)

percentage of their power from nuclear energy include Belgium (54%), Switzerland (38%), Sweden (40%), and South Korea (26%).

Nuclear energy is reliable and emission-free and is viewed by many governments around the world as an attractive form of future electricity generation in the light of controlling greenhouse gas emissions.

3.7.1 Nuclear Challenges

While several new US nuclear plants are planned, none have been built in decades. A new federal law enacted in 2005 seeks to revive the construction and deployment of nuclear power plants by granting regulatory certainty to new and safer designs of nuclear reactors (Fig. 3.8).

The Energy Policy Act of 2005 provides loan guarantees of up to 80% of a project's cost and a production tax credit of 1.8 cents per kWh for new nuclear capacity beginning operation by 2020. The tax credit is specified for the first 8 years of operation and is limited to $125 million per GW per year. This act has spurred applications for new plants to the Nuclear Regulatory Commission. However, their high capital cost is still an impediment to the deployment of these plants.

The Energy Information Administration, an independent agency in the US Department of Energy, assumes the total overnight capital cost for a new nuclear plant is $5366 per kW in 2013 [23] dollars. The EIA has projected the cost of generating electricity from a new nuclear plant in 2020 (i.e., Fig. 3.9) to be 9.52 cents per kWh, about 30% higher than a natural gas combined cycle plant [24].

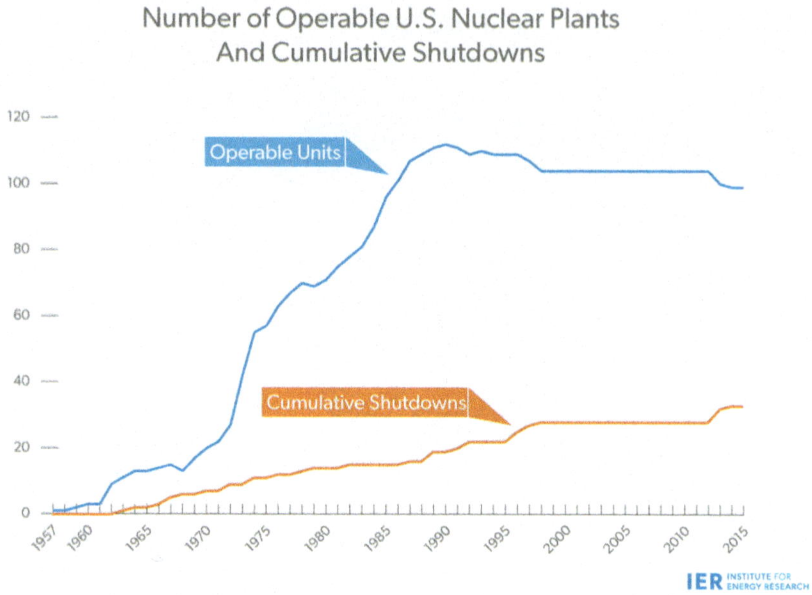

Fig. 3.8 Number of operable US traditional nuclear plants and cumulative shutdowns schedule plan and timeline. (Source: EIA, MER, March 2016)

One ongoing concern in the nuclear power field is the safe disposal and isolation of spent fuel from reactors and waste from reprocessing plants. Consumers have paid billions of dollars in levies on their electric bills over the years to fund a government program to dispose of the waste. The Department of Energy has had the responsibility for the development of waste disposal system for spent nuclear fuel and high-level radioactive nuclear waste since 1982 and plans had called for deposit of the material in Yucca Mountain, Nevada, but funding for this program has essentially been cancelled by the Obama Administration (Fig. 3.10).

Further, the nuclear accident at the Fukushima Daiichi plant in Japan resulting from an earthquake and tsunami in March 2011 has caused nations around the globe to review safety regulations for both existing and new nuclear plants [25].

Almost a year later, in February 2012, the Nuclear Regulatory Commission (NRC) approved a license allowing for the construction and conditional operation of two new nuclear reactors at the Vogtle nuclear power plant in Georgia, marking the first time the commission green-lighted construction for a new reactor since 1978 [26]. Then in March 2012, the NRC approved a second license allowing construction and conditional operation of two new reactors at Scana Corp.'s Virgil C. Summer nuclear power plant in South Carolina [27]. These four nuclear reactors are expected to come on line in 2019 and 2020 [28].

More information on this matter can be found on DOE, NRC, and other related energy government organizations.

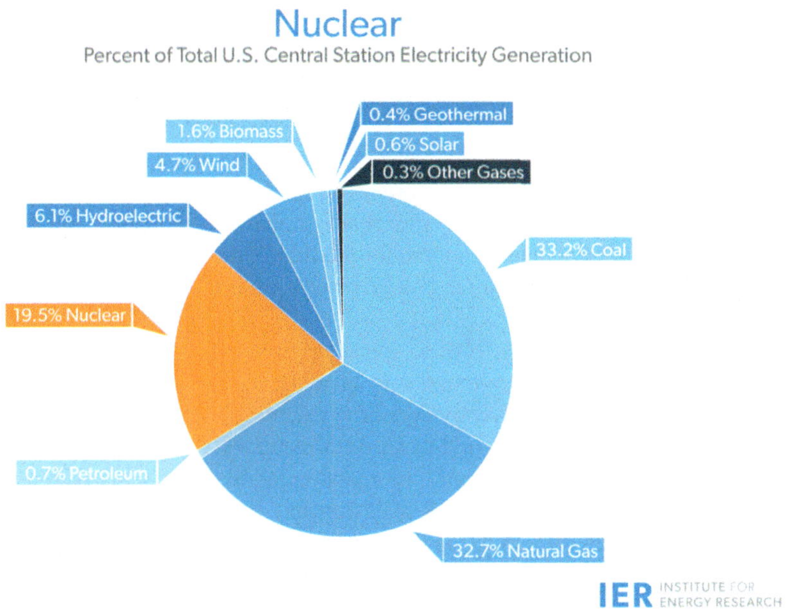

Fig. 3.9 Percent of total US central station electricity generation. (Source: EIA, MER, March 2016)

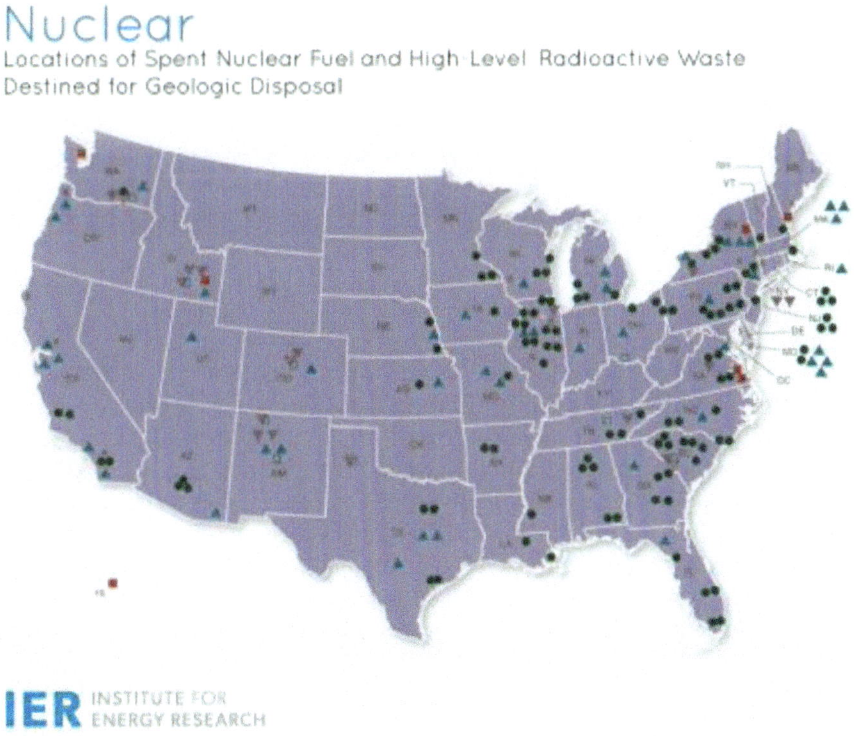

Fig. 3.10 Location of spent nuclear fuel in the USA

3.8 Conclusions

- Unless a number of optimistic assumptions are realized, SMRs are not likely to be a viable solution to the economic and safety problems faced by nuclear power.
- While some SMR proponents are worried that the USA is lagging in the creation of an SMR export market, cutting corners on safety is a shortsighted strategy.
- Since safety and security improvements are critical to establishing the viability of nuclear power as an energy source for the future, the nuclear industry and the DOE should focus on developing safer reactor designs rather than weakening regulations.
- Congress should direct the DOE to spend taxpayer money only on support of technologies that have the potential to provide significantly greater levels of safety and security than currently operating reactors.
- The DOE should not be promoting the idea that SMRs do not require 10-mile emergency planning—nor should it be encouraging the NRC to weaken its other requirements just to facilitate SMR licensing and deployment.
- Department of Defense (DOD) is in need of Mobile Nuclear Micro Reactor for it Brigade deployment and transportability of such reactors.

The Department of Energy (DOE) Office of Nuclear Energy's (NE) Micro Reactor Research, Development, and Deployment (RD&D) Program manages national laboratory-led early-stage generic research and technology development for micro reactor systems and provides cost shared support for micro reactor vendor development and licensing activities through the DOENE Industry Funding Opportunity Announcement. The program also coordinates efforts between the Department of Defense (DOD), industry, and Nuclear Regulatory Commission (NRC) to support the demonstration of micro reactor technology on a DOE national laboratory site. National laboratories supporting the Micro Reactor Program include Idaho National Laboratory (INL), Oak Ridge National Laboratory (ORNL), Los Alamos National Laboratory (LANL), Argonne National Laboratory (ANL), and Sandia National Laboratory (SNL).

As DOE's lead laboratory for nuclear energy, INL has taken the role of the lead laboratory for the micro reactor program. INL has significant historical background as a reactor demonstration location, has been identified as a primary candidate site for micro reactor pilot projects by several reactor vendors, has access to fuel fabrication capabilities, has substantial technical expertise in relevant micro reactor technology development areas, has an extensive nuclear energy R&D infrastructure, and has significant land available for reactor demonstration projects. LANL and ORNL also have significant roles due to their experience in small reactor designs for space applications and advanced materials and manufacturing capabilities. Argonne National Laboratory (ANL) supports advanced micro reactor materials and legacy fuel data qualification while SNL conducts innovative micro reactor energy conversion system R&D. INL closely coordinates the efforts among the participating laboratories.

In FY18, INL and LANL conducted a detailed analyses of commercial and defense micro reactor applications to determine leading high priority micro reactor R&D program areas. These analyses along with DoD and industry stakeholder feedback have formed the basis for FY19 DOE-funded micro reactor efforts in the following areas: accelerating micro reactor High-Assay Low Enriched Uranium (HALEU) production and fuel fabrication capabilities, preparing potential national laboratory micro reactor demonstration sites, demonstrating innovative cross-cutting micro reactor technologies such a heat pipes and advanced moderators, qualifying advanced high temperature materials, exploring additive manufacturing techniques, developing remote monitoring and semiautonomous control systems, and assessing potential DOE, DOD, and NRC regulatory pathways for both near-term micro reactor demonstration licensing and future "nth of a kind" commercial applications.

References

1. https://www.power-technology.com/news/department-energy-national-reactor-innovation-centre/
2. B. Zohuri, P. McDaniel, *Advanced Smaller Modular Reactors: An Innovative Approach to Nuclear Power*, 1st edn. (Springer, Cham, 2019). https://www.springer.com/us/book/9783030236816
3. B. Zohuri, P. McDaniel, *Combined Cycle Driven Efficiency for Next Generation Nuclear Power Plants: An Innovative Design Approach*, 2nd edn. (Springer, Cham, 2018). https://www.springer.com/gp/book/9783319705507
4. B. Zohuri, P. McDaniel, C.R. De Oliveria, Advanced nuclear open air-Brayton cycles for highly efficient power conversion. Nucl. Technol. **192**(1), 48–60 (2015). https://doi.org/10.13182/NT14-42
5. B. Zohuri, *Combined Cycle Driven Efficiency for Next Generation Nuclear Power Plants: An Innovative Design Approach* (201)
6. K. Stacey, Small Modular Reactors are Nuclear Energy's Future, *Financial Times*, 25 July 2016
7. Small Modular Reactors: A Window on Nuclear Energy, Andlinger Center, Princeton University, June 2015
8. B. Zohuri, *Heat Pipe Application in Fission Driven Nuclear Power Plants*, 1st edn. (Springer Publishing Company, Cham, 2019)
9. B. Zohuri, *Heat Pipe Design and Technology: Modern Applications for Practical Thermal Management*, 2nd edn. (Springer Publishing Company, Cham, 2016)
10. https://www.nei.org/resources/letters-filings-comments/part-810-letter-to-congress. Last accessed 12/6/2019
11. https://docs.wixstatic.com/ugd/5b05b3_734bca69ccc6474d949623c853e8be80.pdf. Accessed 12/6/2019 and http://www.nuclearinnovationalliance.org/part810reform
12. B. Zohuri, *Hybrid Energy Systems: Driving Reliable Renewable Sources of Energy Storage*, 1st edn. (Springer Publishing Company, Cham, 2018)
13. E. Moniz, http://energy.mit.edu/news/why-we-still-need-nuclear-power/
14. B. Zohuri, *Neutronic Analysis for Nuclear Reactor Systems* (Springer Publishing Company, Cham, 2016)
15. https://www.uxc.com/p/products/rpt_smo.aspx. Last accessed 12/6/2019

16. https://www.utilitydive.com/news/advanced-us-nukes-need-a-boost-is-the-pentagon-the-answer/559088/
17. Energy Information Administration, Monthly Energy Review, Mar 2016, http://www.eia.gov/totalenergy/data/monthly/pdf/sec1_7.pdf
18. http://www.eia.gov/totalenergy/data/monthly/pdf/sec7_5.pdf
19. Energy Information Administration, http://www.eia.gov/tools/faqs/faq.cfm?id=228&t=21 and Wikipedia, https://en.wikipedia.org/wiki/Watts_Bar_Nuclear_Generating_Station
20. Energy Information Administration, Monthly Energy Review, Mar 2016, http://www.eia.gov/totalenergy/data/monthly/pdf/sec7_5.pdf
21. Energy information Administration, Monthly Energy Review, Table 8.1, Mar 2016, http://www.eia.gov/totalenergy/data/monthly/pdf/sec8_3.pdf
22. Energy Information Administration, Assumptions to the Annual Energy Outlook 2015, Electricity Market Module, http://www.eia.gov/forecasts/aeo/assumptions/pdf/electricity.pdf
23. Energy Information Administration, Assumptions to the Annual Energy Outlook 2015, Table 8.2, http://www.eia.gov/forecasts/aeo/assumptions/pdf/electricity.pdf
24. Energy Information Administration, http://www.eia.gov/forecasts/aeo/electricity_generation.cfm and https://www.instituteforenergyresearch.org/topics/policy/electricity-generation-cost/
25. The New York Times, 21 July 2011, http://www.nytimes.com/2011/07/21/science/earth/21nuke.html?_r=1&nl=todaysheadlines&emc=tha24
26. The Hill, OVERNIGHT ENERGY: Chu to tout nuclear in Georgia, 14 Feb 2012, http://thehill.com/blogs/e2-wire/e2-wire/210651-overnight-energy
27. The Hill, Regulators approve construction of nuclear reactors in South Carolina, 30 Mar 2012, http://thehill.com/blogs/e2-wire/e2-wire/219277-regulators-approve-construction-of-second-new-nuclear-reactors-in-decades?utm_campaign=E2Wire&utm_source=twitterfeed&utm_medium=twitter
28. World Nuclear News, Start date for Vogtle units, 30 Jan 2015, http://www.world-nuclear-news.org/NN-Start-date-delay-for-Vogtle-units-3001158.html and Power Magazine, Costs and Deadlines Continue to Challenge V.C. Summer Nuclear Project, 19 Aug 2015, http://www.powermag.com/challenges-continue-for-summer-nuclear-plant-project/

Index

B. Zohuri, *Nuclear Micro Reactors*, https://doi.org/10.1007/978-3-030-47225-2

The manufacturer's authorised representative in the EU is Springer Nature Customer Service Centre GmbH, Europaplatz 3, 69115 Heidelberg, Germany. If you have any concerns regarding our products, please contact ProductSafety@springernature.com

Printed and bound by CPI Group (UK) Ltd, Croydon, CR0 4YY
30/09/2026
02225471-0007